T0297355

Fatigue in Friction Stir Welding

Fatigue in Friction Stir Welding

J. Brian Jordon
Department of Mechanical Engineering, The University of Alabama,
Tuscaloosa, AL, United States

Harish Rao
National Energy Technology Laboratory,
Department of Energy, Albany, OR, United States

Robert Amaro
Department of Mechanical Engineering, The University of Alabama,
Tuscaloosa, AL, United States

Paul Allison
Department of Mechanical Engineering, The University of Alabama,
Tuscaloosa, AL, United States

Butterworth-Heinemann
An imprint of Elsevier

Butterworth-Heinemann is an imprint of Elsevier
The Boulevard, Langford Lane, Kidlington, Oxford OX5 1GB, United Kingdom
50 Hampshire Street, 5th Floor, Cambridge, MA 02139, United States

Copyright © 2019 Elsevier Inc. All rights reserved.

No part of this publication may be reproduced or transmitted in any form or by any means,
electronic or mechanical, including photocopying, recording, or any information storage and
retrieval system, without permission in writing from the publisher. Details on how to seek
permission, further information about the Publisher's permissions policies and our arrangements
with organizations such as the Copyright Clearance Center and the Copyright Licensing Agency,
can be found at our website: http://www.elsevier.com/permissions.

This book and the individual contributions contained in it are protected under copyright by the
Publisher (other than as may be noted herein).

Notices
Knowledge and best practice in this field are constantly changing. As new research and
experience broaden our understanding, changes in research methods, professional practices, or
medical treatment may become necessary.

Practitioners and researchers must always rely on their own experience and knowledge in
evaluating and using any information, methods, compounds, or experiments described herein. In
using such information or methods they should be mindful of their own safety and the safety of
others, including parties for whom they have a professional responsibility.

To the fullest extent of the law, neither the Publisher nor the authors, contributors, or editors,
assume any liability for any injury and/or damage to persons or property as a matter of products
liability, negligence or otherwise, or from any use or operation of any methods, products,
instructions, or ideas contained in the material herein.

British Library Cataloguing-in-Publication Data
A catalogue record for this book is available from the British Library

Library of Congress Cataloging-in-Publication Data
A catalog record for this book is available from the Library of Congress

ISBN: 978-0-12-816131-9

For Information on all Butterworth-Heinemann publications
visit our website at https://www.elsevier.com/books-and-journals

Working together
to grow libraries in
developing countries

www.elsevier.com • www.bookaid.org

Publisher: Matthew Deans
Acquisition Editor: Christina Gifford
Editorial Project Manager: Ana Claudia A. Garcia
Production Project Manager: Sruthi Satheesh
Cover Designer: MPS

Typeset by MPS Limited, Chennai, India

CONTENTS

This is the ninth volume in this short book series on friction stir welding and processing. As highlighted in the preface of the first book, the intention of this book series is to serve engineers and researchers engaged in advanced and innovative manufacturing techniques. Friction stir welding was invented more than 20 years back as a solid state joining technique. In this period, friction stir welding has found a wide range of applications in joining of aluminum alloys. Although the fundamentals have not kept pace in all aspects, there is a tremendous wealth of information in the large volume of papers published in journals and proceedings. Recent publications of several books and review articles have furthered the dissemination of information.

This book is focused on fatigue in friction stir welded alloys. As the field matures, it is important to review critical properties used for structural applications. Friction stir welding has been successfully implemented for many applications and its technological reach is continuously growing. This book will provide confidence to designers and engineers to consider friction stir welding for fatigue sensitive applications. It will also serve as a resource for researchers dealing with various aspects of friction stir welding. As stated in the previous volumes, this short book series on friction stir welding and processing will include books that advance both the science and technology.

Rajiv S. Mishra
University of North Texas, Denton, TX, United States
February 4, 2019

ACKNOWLEDGMENTS

The authors would like to sincerely thank their students who assisted in data gathering and figure plotting for this book: Conner Cleek, Coleen Fritz, Ben Rutherford, and Robby Escobar. In addition, the authors are grateful of work of current and past students that portions of this book are derived from including Ben White, Dustin Avery, Dr. Abby Cisko, Dr. Joao Moraes, and Dr. Rogie Rodriguez. Additionally, the authors would like to thank Dr. Rajiv Mishra for the invitation to contribute to his FSW book series. Lastly the authors are thankful for the support of their respective families. JBJ expresses gratitude to his wife Amy and his children for their support and encouragement. HMR thanks his wife Pratibha for her constant support. RLA would like to thank his wife, Erika Stinson. PGA would like to give a special thanks to his wife Katherine and his children for their patience and support.

CHAPTER *1*

Introduction to Fatigue in Friction Stir Welding

1.1 INTRODUCTION

Joining of engineering materials continues to be an ongoing challenge in today's manufacturing environment. The challenge is in part due to the use of new advanced material systems or traditionally hard-to-weld alloys. In addition, the desire to join dissimilar materials for light-weighting purposes and the environmental performance requirements has pressed engineers to explore new and innovative joining methods. However, nonferrous alloys, dissimilar joints, and materials with advanced processing histories (i.e., advanced high-strength steels) present significant barriers to traditional fusion joining methods. Although the aerospace-influenced joining methods such as self-pierce riveting have made headlines in commercial vehicle products like the 2015 Ford F-150 light truck, joining by metallurgical bonding in many cases is still the preferred approach. As such, friction stir welding (FSW) overcomes many of the problems associated with fusion welding including solidification cracking, residual stresses, and liquid metal embrittlement. However, as with any new or nontraditional joining techniques, knowledge gaps exist regarding the durability and fatigue performance that may hinder widespread use. In particular, a lack of experimental fatigue data and appropriate design guidelines may cause engineers to consider other joining technologies or more expensive design choices.

1.2 FATIGUE DAMAGE IN ENGINEERING STRUCTURES

A majority of failures of engineering materials and structures are a result of a progressive damage mechanism referred to as fatigue. The mechanism of failure by fatigue is a result of successive repeated loading of applied stresses below the loads necessary for permanent deformation of the material, and in some cases, well below the yield

Fatigue in Friction Stir Welding. DOI: https://doi.org/10.1016/B978-0-12-816131-9.00001-5
© 2019 Elsevier Inc. All rights reserved.

strength of the material. As such, fatigue failure is commonly considered to be an unexpected and catastrophic event, which in some cases results in significant economic and human losses. In particular, catastrophic failure from metal fatigue generally creates international news headlines, for example, when the fatigue failure involves the airline industry, where metal fatigue can lead to fracture of aircraft structures and engine components. In fact, in the early 1950s, de Havilland, a jet airliner manufacturer based in the United Kingdom, never fully recovered from the negative publicity of several fatal crashes of its first commercial passenger jet, the Comet. Failure analysis of the Comet disasters concluded that fatigue cracks initiated from various locations in the fuselage, including poorly designed joints, and grew large enough to cause the aircraft to break apart during flight. Much of the knowledge base and design guidelines that engineers have today did not exist for engineers in the 1950s. Although our understanding of fatigue of materials and structures has grown significantly over the years, failure from metal fatigue is still an ongoing engineering problem. As recently as April 2018, the National Transportation Safety Board blamed metal fatigue as a contributing factor for the engine failure of a Southwest jet airliner which resulted in the death of a passenger [1].

Failure by fatigue in metals is considered the combination of crack incubation and growth mechanisms. In metals for instance, fatigue cracks tend to incubate at microscale discontinuities. These microscale discontinuities can include microstructural features like intermetallic particles or flaws such as casting pores or oxide films. The mechanics of crack propagation and, more importantly, the fatigue crack growth rate is dependent on several factors including the geometry, magnitude of the applied stress, loading type, and microstructural features, and defects. In addition, residual stress and environmental factors can significantly accelerate crack growth and thus reduce the fatigue resistance of the material or component. In welded joints, similar crack incubation and propagation mechanisms exist. However, it is commonly understood that the number of cycles to incubate a fatigue crack in some types of welds is essentially nonexistent due to the presence of weld defects or stress concentrations associated with the weld geometry. As such, much of the life of the material or structure can be spent in the fatigue crack propagation stage. This of course tends to simplify the modeling of fatigue damage by providing a straightforward approach to estimating fatigue life. However, such engineering fatigue models require a robust set of experimental results, and this can be problematic if the weld geometry in the component is significantly different from laboratory test coupons or if the loading case varies substantially. In other cases, welds may exhibit fatigue crack incubation mechanisms similar to parent materials in the high-cycle regime and thus require appropriate experimental and modeling efforts. In addition, variation in weld quality may lead to significant scatter in the number of cycles to failure that further makes estimating the remaining fatigue life difficult without applying a large design factor of safety. Ultimately, the fidelity of the engineering model in predicting the fatigue life of a welded structure depends on a robust understanding of the fatigue behavior and underlining mechanisms, in addition to access of adequate experimental test data.

1.3 BACKGROUND ON FRICTION STIR WELDING

In order to provide context on the topic of fatigue of FSW, a brief review of the FSW process is presented. FSW is a solid-state thermo-mechanical welding process in which a rotating cylindrical weld tool comprising a tool shoulder and probe pin moves along the welding region of the workpiece to be joined. Initially the probe pin is plunged into the weld region of the workpiece, which in turn, generates frictional heat. Upon further plunging the weld tool, additional frictional heat is generated as the rotating tool shoulder comes in contact with the top surface of the workpiece. This heat is sufficient enough to soften the material around the probe pin and under the tool shoulder. The combined action of the probe pin and tool shoulder results in severe plastic deformation and flow of the plasticized metal that occurs as the tool moves along the weld region. In the FSW process, material gets transported from the front of the tool to the trailing edge where the downward force of the tool forges the workpiece. A schematic representation of the FSW in a butt weld configuration is shown in Fig. 1.1 [2].

The weld formed by FSW is generally free of defects that are commonly observed in fusion welding. The FSW process is also free of fumes and does not consume filler materials, and the distortion is typically lower when compared to fusion welding distortion. During the FSW process, the material is removed from the leading edge of the rotating side of the tool, which is in the traversing direction of the tool. This side of the weld is called the advancing side.

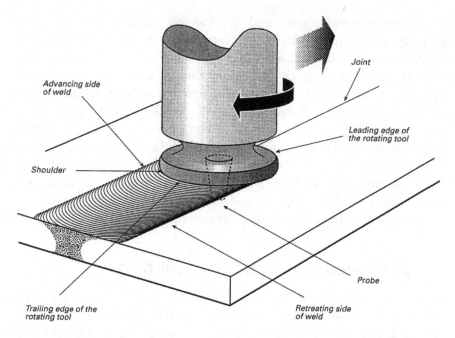

Figure 1.1 Schematic of the friction stir welding (FSW) process [2].

The extracted material is retrieved back into the weld zone at the other end of the rotating tool, which is opposite to the traverse direction and is called the retreating side. Parameters that have a large influence on the structural integrity of the weld include the tool geometry, weld tool rotation rate, weld tool plunge depth, and weld tool traverse speed.

The most common FSW joint type and easiest to fabricate is the butt joint. In fact, NASA uses the butt FSW welds in the fabrication of the new Space Launch System. However, other common types of joints include overlap and T-joints, among others. The fatigue behavior of these and other FSW joints will be discussed in more detail in Chapter 2, Fatigue Behavior in Friction Stir Welds. In aerospace and automotive manufacturing, overlap joints are a convenient method for joining sheet metal together. An example of the use of FSW overlap joints in the automotive industry was the fabrication of the doors first introduced on the 2003 Mazda RX-8 sports car. Although the mechanics of the weld process is similar, FSW overlap in joining of sheet metal is slightly different than the butt configuration. A schematic of the overlap FSW joint and layout for test specimens and clamping fixture is shown in Fig. 1.2 [3].

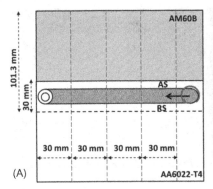 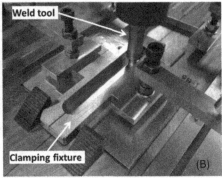

Figure 1.2 (A) An example of the overlap friction stir welding (FSW) methodology indicating the tool traverse direction, tool rotation direction, geometrical dimensions of the overlap test coupons, (B) and an example of a clamping fixture used to weld overlap FSW joints [3].

Similar to the butt joint configuration, in the overlap configuration, the rotating cylindrical weld tool consists of a tool shoulder and probe pin. However, in the overlap configuration, the probe pin penetrates the upper sheet completely and then passes into the bottom sheet to various depths according to the process condition. The downward force and rotational speed of the tool generates frictional heat at the tool–material interface. This frictional heat is sufficient enough to plastically deform metal adjacent to the tool, and a solid-state bond is formed at the interface of the two sheets to be joined. The amount of frictional heat generated and plastic deformation depends on the downward force, dwell time of the tool, and plunge depth of the tool.

The friction stir spot welding (FSSW) process, which is an important and often used variant of FSW that is employed similarly to resistant spot welds, is characterized by a keyhole that is left behind after the extraction of the tool at the end of the weld process. Fig. 1.3A shows a schematic of the overlap FSSW clamping fixture and Fig. 1.3B shows the FSSW joint panel. Unique to both overlap FSW and FSSW joints is the formation of partially bonded "hooks" at the interface of the two sheets. During welding, the trapped oxide films present between the overlapping sheets are often displaced in an upward direction toward the top sheet into a "hook-like" shape. This hooking is largely due to the plastic flow of the material resulting from the downward plunge of the pin into the bottom sheet.

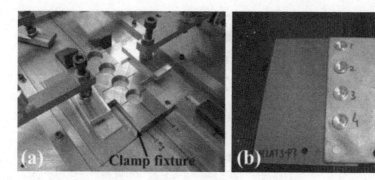

Figure 1.3 (A) Example of a clamping fixture used in overlap friction stir spot welding (FSSW) joints, (B) and completed overlap FSSW panel joint [4].

For successful implementation of overlap FSW and FSSW joints, it is important to understand the mechanisms behind the formation of the weld bond, the failure mechanisms, and the strength of the welds. Various studies have shown that the overlap FSW and FSSW static strength depends in part on the weld geometry, which is highly influenced by the geometry of the weld tool. The geometric features that control strength of the FSW and FSSW overlap joints include the weld bond width, the interfacial "hook" height, and the effective sheet thickness. Of course, process parameters also influence the material flow and thus have a significant impact on the control of defects.

1.4 MOTIVATION AND SUMMARY

As FSW becomes more widely implemented in the joining of structures in manufacturing, design issues related to fatigue performance and the associated mechanisms are of increased importance. Fatigue of FSW is generally a less published topic compared to studies on the microstructure or static mechanical properties. This is largely due to the amount of samples needed to prepare for laboratory testing and also because many OEMs do not make their experimental data available to the public. In any case, published work on the fatigue of different types of FSW under various testing conditions, and material combinations exist in the open literature. Although significant knowledge and expertise exists on the FSW process and the relationship between welding parameters and sound joint fabrication, there is no concise and centralized source of information on the fatigue behavior of FSW. Because FSW joints are fundamentally different than fusion welds, significant knowledge gaps exist regarding fatigue in FSW. As such, the aim of this book is to provide introductory knowledge on the mechanisms of fatigue in FSW joints.

It is intended that the primary audience for this book is the graduate level mechanical, civil, aerospace engineering, and materials science students. However, it is expected that this book can be a reference for fatigue engineers not familiar with FSW, and FSW experts with limited fatigue experience. The first few chapters will provide background necessary to understand FSW and the damage mechanisms associated with fatigue in FSW. These chapters will be presented at a level that first-year graduate students can grasp, and which will provide the foundation for the proceeding chapters. Later chapters will build on the fatigue mechanism foundation by presenting more specific and unique fatigue issues and also provide modeling and design recommendations. The appendix section will present representative fatigue data sets that the reader can use for reference and design purposes.

Regarding specific content in the proceeding chapters, Chapter 2, Fatigue Behavior in Friction Stir Welds, will provide an introduction to fatigue behavior of various joint configurations including butt, overlap, and cross-tension for both linear and spot welds. In addition, an overview of the macroscale behavior of various materials joined with FSW is presented. Chapter 3, Influence of Welding Parameters on Fatigue Behavior, will focus on the influence of welding process parameters on fatigue behavior. In this chapter, the relationship between fatigue and the FSW process window including tool design will be presented. Other factors such as the effect of residual stress and strengthening mechanisms on fatigue performance are discussed briefly. Fatigue Crack growth mechanisms in FSW, including a brief review of fracture mechanics, is the objective of Chapter 4, Fatigue Crack Growth in Friction Stir Welds. Additionally, discussion regarding mean stress, crack closure, and residual stress topics as they pertain to fatigue crack growth in FSW is presented. Chapter 5, Fatigue Modeling of Friction Stir Welding, presents a review of modeling approaches for FSW along with some corresponding examples from literature. The engineering models discussed in this chapter include stress-life, strain-life, structural stress, damage tolerance, and microstructure-sensitive approaches. Chapter 6, Extreme Conditions and Environments, covers extreme conditions and environments that FSW joints may be subjected to under fatigue loading. This chapter will also provide an overview on variable amplitude, multiaxial loading, and the effect of corrosion on fatigue in FSW. The final chapter (see Chapter 7: Beyond Friction Stir Welding: Friction Stir Processing and Additive Manufacturing) presents a brief overview of the fatigue in friction stir processing and additive manufacturing. Lastly, any article or book on the topic of metal fatigue that does not include experimental fatigue results misses an opportunity to provide valuable reference data. As such, an appendix is included in this book that presents experimental fatigue data taken from published literature. This appendix is divided into 4 main sections on fatigue data of FSW joints: (Section 8.1) stress-life; (Section 8.2) strain-life; (Section 8.3) crack growth; (Section 8.4) lap-joints. Each of the sections provides representative data for various types of materials and material combinations.

REFERENCES

[1] Blown Southwest jet engine showed metal fatigue: NTSB. <https://www.cnbc.com/2018/04/18/blown-southwest-jet-engine-showed-metal-fatigue-ntsb.html>; n.d. [accessed January 6, 2018].

[2] Thomas W, Nicholas E. Friction stir welding for the transportation industries. Mater Des 1997;18:269–73. Available from: https://doi.org/10.1016/S0261-3069(97)00062-9.

[3] Rao HM, Ghaffari B, Yuan W, Jordon JB, Badarinarayan H. Effect of process parameters on microstructure and mechanical behaviors of friction stir linear welded aluminum to magnesium. Mater Sci Eng A 2016;651. Available from: https://doi.org/10.1016/j.msea.2015.10.082.

[4] Rao HM, Yuan W, Badarinarayan H. Effect of process parameters on mechanical properties of friction stir spot welded magnesium to aluminum alloys. Mater Des 2015;66:235–45. Available from: https://doi.org/10.1016/J.MATDES.2014.10.065.

CHAPTER 2

Fatigue Behavior in Friction Stir Welds

2.1 INTRODUCTION

Due to the energy efficiency requirements, economic feasibility, and versatility to join a diverse group of materials, friction stir welding (FSW) has found wide spread applications in recent years. Particularly, the automotive and aerospace industry has implemented FSW processes to join a wide variety of structural materials [1]. Because most of these FSW joints are designed for applications to withstand cyclic loading during the service life of the structure, it is essential to develop a robust understanding of the factors influencing the fatigue behavior of FSW joints. As a starting point to designing against failure from fatigue, laboratory fatigue tests of the representative FSW joint aid in understanding the ability of the FSW to withstand the progressive damage and ultimate fracture when subjected to cyclic loading under various environmental conditions. Similar to other welds, FSW joints under fatigue loading exhibit distinct fracture mechanisms that are unique based on the loading type and joint configuration. Because there are multiple factors that can directly influence the fatigue performance, it is imperative to develop a comprehensive understanding of the fatigue behavior in a structural joint fabricated using FSW.

Because design against fatigue failure begins with understanding fatigue mechanisms, which in many cases begins in the laboratory where variables can be controlled, this chapter begins with a review of joint configurations of FSW. Regarding laboratory fatigue testing, a brief discussion of common fatigue tests for various coupon types is also introduced. To provide the reader with appropriate background on fracture of FSW subjected to fatigue loading, macroscale fracture mechanisms of FSW coupons subjected to fatigue is discussed. Finally, a brief summary on general fatigue behavior relative to specific materials is presented. Please note that fatigue crack growth testing and analysis is treated as a separate topic and thus, the reader is directed to Chapter 4, Fatigue Crack Growth in Friction Stir Welds.

Fatigue in Friction Stir Welding. DOI: https://doi.org/10.1016/B978-0-12-816131-9.00002-7
© 2019 Elsevier Inc. All rights reserved.

2.2 WELD TYPES AND JOINT CONFIGURATIONS

The FSW process is widely used to produce spot welds, linear seam welds, and butt welds based on the material thickness and design requirements. Friction stir spot welding (FSSW) is ordinarily used for producing spot welds in overlap sheet metals as shown in Fig. 2.1A [2]. In spot welds, the weld tool is plunged on to the top sheet to a controlled depth, whereas the tool pin completely penetrates the top sheet and partially penetrates the bottom overlapping sheet. The frictional heat generated between the tool shoulder and top sheet, along with the stirring action of the tool pin in the bottom sheet, plastically deforms the overlapping sheets to form a solid-state bond. In friction stir linear welding (FSLW) however, the weld tool after plunging is traversed for a predetermined length along the faying surface to obtain a continuous linear weld as shown in Fig. 2.1B [3]. FSLW is commonly used to produce welds in overlap and butt weld joints. Universally, spot welds are a default choice for producing joints in thin-section materials like sheet metals in an overlap configuration. Although linear welds are a default choice used to obtain butt joints in plate materials, linear welds are also used to obtain joints in the overlap configuration based on specific

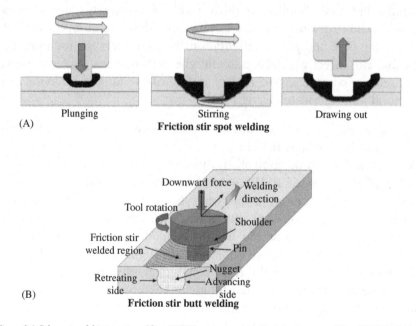

Figure 2.1 Schematic of friction stir welding (FSW) processes. (A) friction stir spot welding (FSSW) [2] and (B) friction stir linear weld in butt joint configuration (FSBW) [3].

design considerations. Henceforth, all lap-shear FSLW will be addressed as FSLW and all FSW butt welds be addressed as FSBW.

As a design engineer, understanding the fatigue life and fracture criteria for a given weld type and material combination is essential before that joint type can be integrated into the final design of a structure. In early stages of design, once the material is chosen, the weld type is determined based on the loads/forces the structure would experience during its entire service life. Estimation of fatigue life and the development of fracture criteria are generally accomplished by extensive laboratory testing in a controlled environment. The schematics in Fig. 2.2 presents the typical standard joint configurations that are commonly used in laboratories to estimate the fatigue behavior of FSW joints and to develop fracture criteria for a given material combination and weld type. In FSW, the fracture criteria are defined by the distinctive macro features of the weld which nucleate and propagate the dominant fatigue crack leading to final fracture. Unlike in commonly practiced welding processes like resistance spot welding, where the fracture criteria is defined by the weld nugget diameter or sheet thickness, in FSW no such singular criteria exists. The macro features are very distinctive for a given weld type, material combination, and welding process

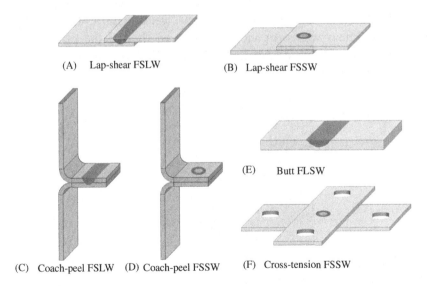

(A)   Lap-shear FSLW   (B) Lap-shear FSSW

(E)   Butt FLSW

(C)   Coach-peel FSLW   (D) Coach-peel FSSW   (F)   Cross-tension FSSW

Figure 2.2 Schematic of commonly used geometric configuration for studying fatigue performance and fracture criteria in spot and linear welds: (A) lap-shear FSLW, (B) lap-shear FSSW, (C) coach-peel FSLW, (D) coach-peel FSSW, (E) FSBW, and (F) cross-tension FSSW.

condition. Therefore, in FSW joints, in addition to estimating fatigue life, understanding the fracture criteria for a given weld and joint type is necessary. In the next section, we will discuss the various FSW joint types that are commonly used for various structural applications.

2.2.1 Lap-Shear Joints

Lap-shear welds are the most commonly used overlap weld joint type in welding sheet metals or thin section materials. This configuration is also one of the most commonly used joint types to study the behavior of the welds/joints under predominately shear forces, in addition to secondary bending forces, as described for both in Fig. 2.3A. Although the shear force primarily acts along the weld nugget, the bending forces resulting from the specimen geometry and loading condition impart a moment on the top and bottom sheets (shown left to right in Fig. 2.3A) comprising the specimen, as indicated in Fig. 2.3A. In a laboratory fatigue test, lap-shear welds are typically tested in a tension−tension fatigue load ($R \geq 0$). Although there are no standard geometries for lap-shear test specimens, as a rule of thumb, the width of test coupon is typically controlled by the length of weld joint overlap as illustrated in Fig. 2.3B by use of the dimensions marked "X." In FSW, lap-shear joints can be either produced by FSSW or by FSLW based on the engineering design requirements. It is important to note that the fatigue tests results from spot welds cannot be compared directly to linear welds (or vice versa), even though the material and overlap area may be identical. This is primarily because both FSSW

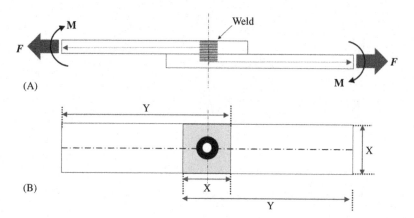

Figure 2.3 Schematic illustrating (A) the shear forces and bending moments in a lap-shear weld and (B) example of a lap-shear fatigue test specimen specifying the geometrical constants.

and FSLW joints exhibit unique macro features and behave differently under cyclic loading. In FSSW lap-shear joints, the fatigue life and fracture mode are mainly influenced by the interfacial hook geometry, effective sheet thickness, and the weld bond width. In FSLW lap-shear joints, the weld bond width and effective sheet thickness have primary influence upon the fatigue life. Additionally, the placement of the advancing side hook (macro feature in FSLW) with respect to load direction has been observed to effect fatigue performance [4].

2.2.2 Cross-Tension and Coach-Peel Joints

In addition to lap-shear specimens, cross-tension and coach-peel joints are also common overlap geometries used in laboratory tests. These geometries are ideal when characterizing welds under specific loading conditions. Cross-tension joints are designed to characterize the weld strength when forces/loads are applied normal to the weld interface in pure tension, as shown in Fig. 2.4A. The cross-tension specimens are primarily produced by FSSW. It should be noted that, due to the geometry of the joint, cross-tension specimens require a specifically designed fixture to hold them between the grips during fatigue testing.

Coach-peel joints are primarily used to characterize the peel strength of welds. The joining of thin-section materials, such as sheet metals, predominantly creates this type of weld specimens. Coach-peel joints can be produced either by FSSW or FSLW. The design of the

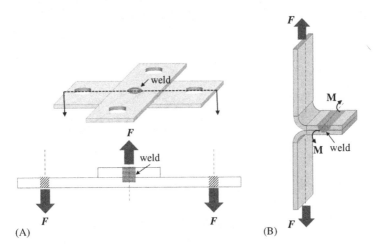

Figure 2.4 Schematic illustrating the forces and moments acting on the weld in (A) a cross-tension test specimen and (B) coach-peel test specimen.

joint results in an introduction of bending moments at the two overlapping sheets, as illustrated in Fig. 2.4B. Cross-tension and coach-peel FSSW joints exhibit similar macro features to those observed in a lap-shear FSSW. Additionally, the coach-peel FSLW joints exhibit similar macro features and microstructure to those observed in a lap-shear FSLW. Given the vastly different manifestation of normal forces, shears, and moments in lap-shear, coach-peel, and cross-tension joints, the fatigue lives of each specimen type are not comparable to those of the other specimen types.

2.2.3 Butt Joints

Butt joints are extensively used in welding relatively thick section materials like plates or hollow cylindrical structures (e.g., pipes and pressure vessels). Butt welds are the most commonly used weld type employed to join structures where the primary load condition is either tension or compression. Once the butt welds are fabricated, the fatigue test specimens are machined to a dog-bone/hour glass shape as illustrated in Fig. 2.5 [5]. Schematics in Fig. 2.6A and B [6] provide examples of (A) constant curvature and (B) wide gage length FSBW fatigue test specimens commonly used in a laboratory fatigue testing. Unlike overlap joints, butt welds can be either fatigue tested in load-control or strain-control mode, in pure tension–tension or tension–compression, based on the engineering design requirements. Due to the nature of the joint, butt welds primarily experience pure tension or compression loads, and in most cases the bending forces are negligible.

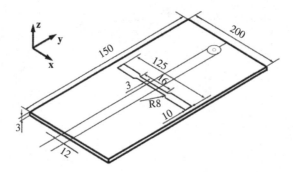

Figure 2.5 Schematic illustration of FSBW AZ91D plates showing the fatigue specimen machining geometry [5]. FSBW, butt-friction stir welding.

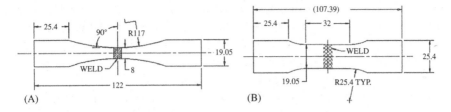

Figure 2.6 (A and B) Geometric outline of commonly used constant curvature and wide gage FSBW fatigue test specimens [6] (all dimensions in millimeters). Reprinted with permission from Springer Nature.

2.3 FATIGUE TEST METHODS

Laboratory fatigue tests are generally performed to evaluate the endurance or life span of engineering materials under cyclic loading. In addition to endurance limit, fatigue tests also provide valuable information on the crack resistance and fracture modes in engineering materials. Today, fatigue tests are typically performed on servo-hydraulic load frames with the vast majority of them conducted at ambient conditions. Fatigue tests can also be performed under (a) elevated or high temperatures to study the creep-fatigue and thermo-mechanical fatigue properties, (b) fretting fatigue conditions when engineering materials are subjected to sliding forces, and (c) corrosion fatigue if engineering materials are required to operate in aggressive chemical environments. In all of the different fatigue tests mentioned previously, the common feature is that the engineering material is subjected to cyclic loading irrespective of the external environment.

There are several basic forms of laboratory controlled fatigue testing: load-controlled fatigue testing and strain-controlled fatigue testing. Generally, the load-controlled fatigue test is performed to evaluate the endurance limit of a given engineering material and the applied cyclic loads are within the elastic load limit of the material. Strain-controlled fatigue tests may be performed to evaluate engineering materials that are exposed to cyclic loads below and beyond their elastic load limit. In addition to uniaxial loading, multi-axial fatigue tests are also performed based upon the loading conditions the engineering materials are designed for during service. As mentioned in the introduction of this chapter, fatigue crack growth testing and analysis is discussed in Chapter 4, Fatigue Crack Growth in Friction Stir Welds.

Generally, FSW overlap joints are tested in load-control mode, whereas the FSBW are tested in load-controlled or strain-controlled cyclic testing. Strain-controlled fatigue tests require extensometers that are attached to test specimens throughout the test and are generally tested according to ASTM E606/E606M-12 standards. When performing the fatigue tests extreme care must be taken while setting up the test specimen between the grips of the hydraulic test frame. The load cell must be calibrated before testing and alignment of the grips must be ensured to avoid introduction of external bending/shear forces acting on the specimen, especially in low-cycle fatigue applications with large strain or stress amplitudes. In addition, shims must be employed for overlap welds to eliminate excessive bending forces during gripping of the offset grip-section for the specimen. The schematic in Fig. 2.7 shows the various types of shims that must be employed for commonly used overlap joint configurations. If the two overlapping sheets are of similar thickness, shims must be placed on the opposite ends of sheets between the jaws of the upper and lower grips as shown in Fig. 2.7A. Similarly, if the overlapping sheets are of different thickness, different sized shims must be placed as shown in Fig. 2.7B to ensure the applied load is along the centerline of the overlap joint. Coach-peel specimens have a tendency to bend during the testing due to the geometry of the joint. Therefore, care must be taken such that the length of the specimen that is inserted between the jaws is equal to the length of the leg just before it curves as indicated in Fig. 2.7C. In cases where the width of the overlapping sheets is much larger or smaller than the jaws of the grips employed, care must be taken such that the center of the sheet/weld aligns with the center of the grips, as shown in Fig. 2.7D for shorter width sheet and Fig. 2.7E for oversized sheet width. Ensuring the proper setup during the fatigue test is extremely important to obtain accurate results and eliminate unintended forces/loads applied on the specimens during fatigue testing.

For cross-tension specimens, a uniquely designed fixture is typically used to hold the specimen between the grips during the fatigue test. Shims are not needed in cross-tension specimens, but welds must be produced at the center of the overlapping sheets. A schematic of the unique cross-tension fixture and a close-up of the fixture during testing are shown in Fig. 2.8A and B, respectively [7]. For the butt welds, test specimens are machined to dog-bone/hour glass shape, and hence care must be taken to properly align the test specimen along the centerline of the grips. Generally, no shims are employed while testing butt welds; however, an extensometer may be used if the fatigue test is performed in strain-controlled mode.

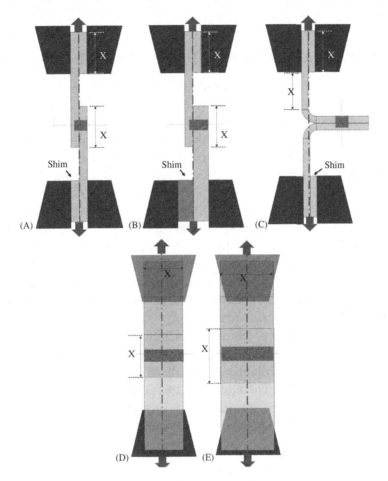

Figure 2.7 Shim placement for various sheet thickness and joint configuration in (A) lap-shear specimens with similar sheet thickness, (B) lap-shear specimens with dissimilar sheet thickness, (C) coach-peel specimens with similar sheet thickness and, (D) and (E) specimen alignment along the centerline of the jaws in undersized and oversized specimens.

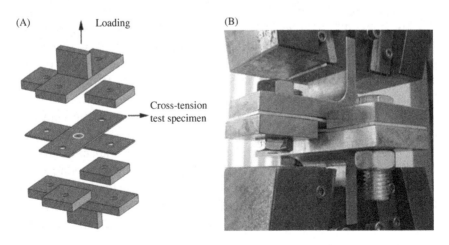

Figure 2.8 (A) Schematic illustration of cross-tension test and (B) A close-up view of the fixture for cross-tension test [7].

2.4 MACRO FEATURES AND FATIGUE BEHAVIOR

FSW is a thermo-mechanical process where the weld joint is obtained by severe plastic deformation of the bulk material at elevated temperatures (but below solidus temperature) and pressure. This unique solid-state joining process introduces several distinctive physical/macro features, which later becomes a permanent part of the weld. Macro features in FSW joints play a crucial role in determining the fatigue life and fracture modes in a given FSW joint. Due to the differences in weld types (FSSW vs FSLW vs FSBW), macro features in FSW joints vary significantly. Hence, understanding the unique macro features for a given weld type and how they affect the fatigue life and fracture mechanisms is essential part of the design process. In addition to macro features, the microstructure of the weld, notably those micro-features in the stir zone and heat-affected zone (HAZ) have a large impact on the fatigue performance in many FSBW joints. Furthermore, the macro features and microstructure may be vastly dependent upon the welding process conditions/parameters, such as weld tool design, tool plunge depth, dwell time, tool traverse speed, and material stacking. Therefore, the impact of process conditions on macro features will be discussed in further detail in Chapter 3, Influence of Welding Parameters on Fatigue Behavior.

2.4.1 Friction Stir Spot Welds

FSSW are generally used when joining thin overlapping sheets or sheet materials. It is one of the most commonly used friction stir welds in the automotive industry due to its ability to weld sheets in short periods of time. In addition to the "keyhole" feature, FSSW are characterized by three unique macro features: the interfacial hooks, effective sheet thickness, and weld bond width. Each of these macro features have been shown to influence the static strength and fatigue performance of the FSSW joints. The schematic in Fig. 2.9 represents the various macro features commonly observed in a FSSW joint. Although interfacial hooks are generally considered as a metallurgical defect and a preexisting crack in the weld, effective sheet thickness and weld bond width are two crucial geometrical features that have a direct impact on the fatigue behavior of a FSSW joint. The weld bond width is a completely metallurgically bonded area and the load-bearing feature in an FSSW joint. Lastly, effective sheet thickness is a purely geometrical entity, where it is the distance measured between the tip of the

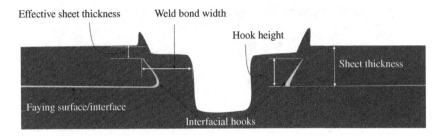

Figure 2.9 Schematic representation of the various macro features in an FSSW joint.

interfacial hook to the nearest free surface of the top overlapping sheet. Laboratory fatigue tests have overwhelmingly shown that these macro features are the dominant factors influencing the fatigue performance and fracture modes in FSSW joints.

Two commonly observed fatigue fracture modes in a FSSW joint are kinked crack growth through the interfacial hooks and weld nugget shear fracture. The interfacial hooks in the weldment act as a preexisting crack and facilitates the initiation of the dominant fatigue crack [2,8−12], which leads to ultimate fracture of the FSSW joint. Under low cyclic loads (high-cycle fatigue regime), the dominant fatigue crack propagates through the effective sheet thickness resulting in fracture and separation of the overlapping sheet from the weld nugget, as seen in Fig. 2.10 [2]. At higher fatigue loads (low-cycle fatigue regime), FSSW joints tend to fail due to interfacial fracture of the weld nugget. The dominant fatigue crack nucleates at the notch-like opening formed due to overlapping of the sheets (faying surface) and propagates through the weld bond width leading to final fracture of the FSSW as seen in Fig. 2.11 [10].

2.4.2 Friction Stir Linear Welds

Macro features in overlap FSLW differ widely from those observed in overlap FSSW joint due to the inherent difference in the welding process. A schematic in Fig. 2.12 represents the various macro features commonly observed in an overlap FSLW joint. In lap-shear FSLW, the importance of advancing side and retreating side hook placement with respect to loading direction has a dominant effect on fatigue performance [4]. During FSLW, the material under the weld tool shoulder is swept upwards, which forms a hook-like feature on the advancing side of the weld tool. This material is retrieved back into the weldment

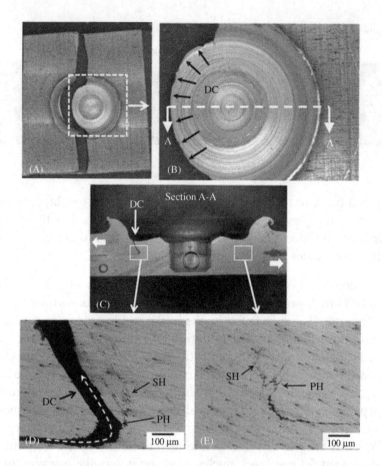

Figure 2.10 Dominant fatigue crack (DC) propagating through the effective sheet thickness and fracture of top sheet in FSSW magnesium AZ31 alloy [2]. (A) An overview of the fractured surface of the top sheet and (B) magnified view of the fractured surface around the keyhole; (C) cross section view of the weld nugget shows the dominant crack (DC) propagating from the weld interface; (D) a high magnification image shows the dominant crack (DC) propagating from the primary hook (PH); and (E) high magnification image of the primary (PH) and the secondary (SH) hooks with no evidence of cracks. Bold arrows in (C) indicate direction of applied loading.

and forms a cold lap feature or otherwise known as the retreating side hook. The advancing side hook is swept backwards and terminates outside the stir zone as shown in Fig. 2.12, whereas the retreating side hook normally terminates inside the stir zone facing away from the loading direction.

Based on the tool rotation, tool traverse direction, and material stacking of the joint, the advancing side hook may form on the free end of the overlapping sheet or may form on the loading side of the

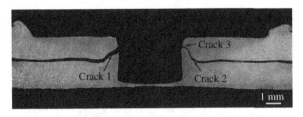

Figure 2.11 Representative interfacial fracture of the weld nugget in FSSW AA6061-T6 alloys [10].

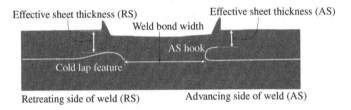

Figure 2.12 Schematic illustrating the various macro features in a typical overlap FSLW joint.

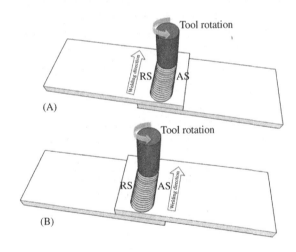

Figure 2.13 Coupon layout of friction stir linear welding (A) retreading side, and (B) advancing side [4].

overlapping sheet as illustrated in Fig. 2.13A and B [4]. Due to the orientation and geometry of the advancing side hook, the fatigue crack will typically propagate at a much higher rate in FSLW lap-shear joints produced with an advancing side setup as illustrated in Fig. 2.13B. Therefore, when designing the FSLW process in a lap-shear joint, it is very important that the welds are produced with

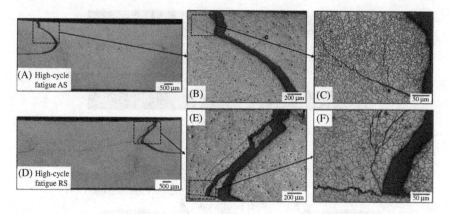

Figure 2.14 Representative cross-sectional views of fractured high-cycle fatigue coupons loaded on (A) advancing side (failure at 353,589 cycles), (D) retreating side (failure at 437,661 cycles). Magnified view of crack path that grew between the stir zone and the thermo-mechanically affected zone: (B) advancing side, (E) retreating side. High magnified view of secondary cracks: (C) advancing side, (F) retreating side. Load ratio was R = 0.1 [4].

advancing side hook oriented as illustrated in Fig. 2.13A. In addition to the hook placement, the weld bond width and the effective sheet thickness can also have an effect on the fatigue performance. Similar to FSSW, the larger the effective sheet thickness and weld bond width are in size, an increase in fatigue life is generally observed. In laboratory fatigue tests, the most commonly observed fracture mode includes the sheet separation due to kinked crack growth from one of the interfacial hooks [4,13–15] based on the hook orientation, as shown in Fig. 2.12. Once the fatigue crack nucleates at one of the hooks, the dominant fatigue crack may propagate through the effective sheet thickness if the hook is oriented towards the top sheet, as shown in Fig. 2.14A (advanced side) and Fig. 2.14D (retreating side). Similarly, if the hook terminates or orients towards the stir zone, then the fatigue crack may propagate through the weld bond width leading to interfacial fracture of the weld.

2.4.3 Friction Stir Butt Welds

Butt welds are a commonly used joint type in welding cylindrical structures, thick section materials, or multi-hollow extrusions where superior weld strength is expected. Butt welds produced using FSW generally exhibit less distortion and a larger weld bond, which significantly improves the mechanical performance of the welds. Unlike overlap FSLW joints, FSBW joints do not exhibit typical macro features like interfacial hooks, but rather the microstructure and defects of the weldment have been observed to primarily influence the fatigue performance. In laboratory fatigue testing, fracture has been most commonly observed to occur in the HAZ of the weld where the grains were predominantly larger and softer compared to base metal or the stir zone. In particular, similar to monotonic tensile loading, fracture usually occurs in the HAZ on the advancing side of the FSBW joint. During the FSBW process, the material on the advancing side of the weld tool provides higher resistance to rotation of the tool which introduces higher frictional heat resulting in marginally softer HAZ compared to the HAZ on the retreading side of the weld [16–18]. This difference in softness of the materials on the advancing side of the weld tool can (but not always) lead to fracture on this side of the weld joint during cyclic loading. The stir zone/weld nugget of a FSBW joint is generally characterized by fine recrystallized grains due to extensive stirring action of the weld tool pin. In some FSBW, it is possible that the weld nugget may have voids or defects due to lack of weld tool penetration or improper weld process parameters. These defects can drastically reduce the fatigue life and lead to fracture of the weld nugget regardless of the orientation of the advancing and retreating sides [19]. Similarly, surface roughness or micro defects on the surface of the weldment as a result of tool shoulder interaction with the top surface of the material during welding can also initiate fatigue cracks leading to fracture. Fig. 2.15 shows a set of representative fractured FSBW specimens in a laboratory fatigue test environment.

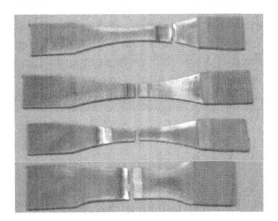

Figure 2.15 Representative of fractured FSBW specimen in a laboratory fatigue test [6]. FSBW, butt-friction stir welding. Reprinted with permission from Springer Nature.

2.5 ALLOYS

Originally, the FSW process was developed predominantly to join aluminum alloys. Over the years, however, the application of FSW spread beyond aluminum alloys to other materials including steels, magnesium alloys, titanium alloys, and polymer composites to a certain extent. As such, a brief summary of fatigue in FSW of various alloys is presented in this section.

2.5.1 Aluminum Alloys

One of the most common materials joined by FSW are aluminum alloys. Specifically, Aluminum 2XXX [20], 5XXX [21,22], and 6XXX [21,23–25] series alloys joined through FSW have generally shown higher fatigue life and an improved endurance limit when compared to welds produced using conventional fusion welding techniques. Similar observations were made when the fatigue behavior of defect-free FSW joints were compared to design codes for fusion welds [20–22,26]. In addition, the scatter in the number of cycles to failure in FSW joint has been reported to be less than those observed in fusion welds [27]. Of course, the quality of the FSW does impact the fatigue life. Dickerson and Przydatek [21] examined the FSW with and without defects and found defect-free FSW joints significantly outperformed conventional fusion welds. However, when defects were present in the FSW joints, the fatigue life failed to meet the design code recommendations. In 7xxx aluminum alloys, joining by fusion welding is difficult due to solidification cracking. However, 7xxx aluminum alloys can be more easily joined with FSW, and fatigue lives of FSW of AA7075 have been reported to exceed design code recommendations [28]. In some non-heat treatable 1XXX-O and 5XXX-O aluminum alloys, the fatigue performance of FSW joints was comparable to the base material. Although in heat treatable alloys, dynamic aging may occur in the stir zone and the thermo-mechanically affected zone which may reduce the fatigue life [17].

2.5.2 Magnesium Alloys

Magnesium alloys are well known for their high strength-to-weight ratios. The inherent challenge that still exists with these alloys is the absence of a robust joining technique. Magnesium alloys are difficult to join using fusion welding techniques and generally results in poor, brittle welds. In recent years, FSW has been extensively used for joining a spectrum of magnesium alloys. In addition to producing superior quality welds when compared to arc welding techniques, FSW has been observed to alter the microstructure of several magnesium alloys. Most of the microstructure changes are in the form of grain refinement, introduction of twinning, and the breaking up of coarse intermetallic particles into finer particles. In cast magnesium alloys like AZ91, FSW breaks up coarse as-cast microstructure and produces a refined grain structure in the weld nugget. FSW can also eliminate casting defects and increase hardness, thereby improving the fatigue performance of the FSW joint [5,29]. Compared to fusion welding, the fatigue performance of AZ31 Mg alloys joined with FSW have been reported to exceed welds joined with arc welding [30].

2.5.3 Steel and Other Hard Alloys

FSW of "hard" alloys like steel and titanium is challenging due to the limitations of the weld tooling material. FSW of hard alloys requires very high strength weld tools to produce sufficient frictional heat required for metallurgical bonding. Additionally, FSW of steels leads to significantly higher rates of tool wear compared to aluminum and magnesium alloys. Therefore, the weld tools are generally made, or coated, with much harder materials than that being welded. In fact, coatings like carbides have been used to extend the weld tool life. Steel alloys that are joined by fusion welding typically exhibit large amounts of distortion and have the potential to introduce residual stress that can significantly reduce the fatigue life of the welded joints [31–33]. In some interstitial free steels, the FSW process refined the coarse-grain structure of the base metal to fine nanograins, which resulted in an increased fatigue life [34]. FSW steel alloys may fracture in the base metal during quasi-static testing thus signaling an overmatched weld. However, this characteristic of steel FSW should not be considered as a reference for weld quality, but rather a thorough investigation of the weld joint has to be performed under cyclic loading to correctly identify the weld quality. Often times, the internal flaws in the weld nugget or the surface flaws may not contribute much to fracture in quasi-static loading but may significantly impact the fatigue performance. FSW can also significantly modify the microstructure of the base metal, as in the case of AISI 409M ferritic steel, by producing dual ferritic–martensitic microstructure [33].

REFERENCES

[1] Ma ZY. Friction stir processing technology: a review. Metall Mater Trans A 2008;39:642–58. Available from: https://doi.org/10.1007/s11661-007-9459-0.

[2] Rao HM, Jordon JB, Barkey ME, Guo YB, Su X, Badarinarayan H. Influence of structural integrity on fatigue behavior of friction stir spot welded AZ31 Mg alloy. Mater Sci Eng A 2013;564:369–80. Available from: https://doi.org/10.1016/j.msea.2012.11.076.

[3] Mishra RS, Ma ZY. Friction stir welding and processing. Mater Sci Eng R Reports 2005;50:1–78. Available from: https://doi.org/10.1016/j.mser.2005.07.001.

[4] Moraes JFCFC, Rodriguez RII, Jordon JBB, Su X. Effect of overlap orientation on fatigue behavior in friction stir linear welds of magnesium alloy sheets. Int J Fatigue 2017;100:1–11. Available from: https://doi.org/10.1016/j.ijfatigue.2017.02.018.

[5] Ni DR, Chen DL, Yang J, Ma ZY. Low cycle fatigue properties of friction stir welded joints of a semi-solid processed AZ91D magnesium alloy. Mater Des 2014;56:1–8. Available from: https://doi.org/10.1016/j.matdes.2013.10.081.

[6] Garware M, Kridli GT, Mallick PK. Tensile and fatigue behavior of friction-stir welded tailor-welded blank of aluminum alloy 5754. J Mater Eng Perform 2010;19:1161–71. Available from: https://doi.org/10.1007/s11665-009-9589-1.

[7] Shen Z, Yang X, Yang S, Zhang Z, Yin Y. Microstructure and mechanical properties of friction spot welded 6061-T4 aluminum alloy. Mater Des 2014;54:766–78. Available from: https://doi.org/10.1016/j.matdes.2013.08.021.

[8] Tran V-X, Pan J, Pan T. Fatigue behavior of aluminum 5754-O and 6111-T4 spot friction welds in lap-shear specimens. Int J Fatigue 2008;30:2175–90. Available from: https://doi.org/10.1016/j.ijfatigue.2008.05.025.

[9] Su ZM, He RY, Lin PC, Dong K. Fatigue of alclad AA2024-T3 swept friction stir spot welds in cross-tension specimens. J Mater Process Technol 2016;236:162–75. Available from: https://doi.org/10.1016/j.jmatprotec.2016.05.014.

[10] Venukumar S, Muthukumaran S, Yalagi SG, Kailas SV. Failure modes and fatigue behavior of conventional and refilled friction stir spot welds in AA 6061-T6 sheets. Int J Fatigue 2014;61:93–100. Available from: https://doi.org/10.1016/j.ijfatigue.2013.12.009.

[11] Hong SH, Sung S-J, Pan J. Failure mode and fatigue behavior of dissimilar friction stir spot welds in lap-shear specimens of transformation-induced plasticity steel and hot-stamped boron steel sheets. J Manuf Sci Eng 2015;137:051023. Available from: https://doi.org/10.1115/1.4031235.

[12] Jordon JB, Horstemeyer MF, Daniewicz SR, Badarinarayan H, Grantham J. Fatigue characterization and modeling of friction stir spot welds in magnesium AZ31 alloy. J Eng Mater Technol 2010;132:041008. Available from: https://doi.org/10.1115/1.4002330.

[13] Fersini D, Pirondi A. Fatigue behaviour of Al2024-T3 friction stir welded lap joints. Eng Fract Mech 2007;74:468–80. Available from: https://doi.org/10.1016/j.engfracmech.2006.07.010.

[14] Rao HM, Jordon JB, Ghaffari B, Su X, Khosrovaneh AK, Barkey ME, et al. Fatigue and fracture of friction stir linear welded dissimilar aluminum-to-magnesium alloys. Int J Fatigue 2016;82:737–47.

[15] Costa MI, Leitão C, Rodrigues DM. Influence of post-welding heat-treatment on the monotonic and fatigue strength of 6082-T6 friction stir lap welds. J Mater Process Technol 2017;250:289–96. Available from: https://doi.org/10.1016/j.jmatprotec.2017.07.030.

[16] Lemmen HJK, Alderliesten RC, Benedictus R. Evaluating the fatigue initiation location in friction stir welded AA2024-T3 joints. Int J Fatigue 2011;33:466–76. Available from: https://doi.org/10.1016/j.ijfatigue.2010.10.002.

[17] Uematsu Y, Tokaji K, Shibata H, Tozaki Y, Ohmune T. Fatigue behaviour of friction stir welds without neither welding flash nor flaw in several aluminium alloys. Int J Fatigue 2009;31:1443−53. Available from: https://doi.org/10.1016/j.ijfatigue.2009.06.015.

[18] Besel M, Besel Y, Alfaro Mercado U, Kakiuchi T, Uematsu Y. Fatigue behavior of friction stir welded Al-Mg-Sc alloy. Int J Fatigue 2015;77:1−11. Available from: https://doi.org/10.1016/j.ijfatigue.2015.02.013.

[19] Zhou C, Yang X, Luan G. Effect of root flaws on the fatigue property of friction stir welds in 2024-T3 aluminum alloys. Mater Sci Eng A 2006;418:155−60. Available from: https://doi.org/10.1016/j.msea.2005.11.042.

[20] Di SS, Yang XQ, Luan GH, Jian B. Comparative study on fatigue properties between AA2024-T4 friction stir welds and base materials. Mater Sci Eng a-Structural Mater Prop Microstruct Process 2006;.

[21] Dickerson TL, Przydatek J. Fatigue of friction stir welds in aluminium alloys that contain root flaws. Int J Fatigue 2003;25:1399−409. Available from: https://doi.org/10.1016/S0142-1123(03)00060-4.

[22] Lomolino S, Tovo R, Dos Santos J. On the fatigue behaviour and design curves of friction stir butt-welded Al alloys. Int J Fatigue 2005;27:305−16. Available from: https://doi.org/10.1016/j.ijfatigue.2004.06.013.

[23] Ranes M, Kluken AO, Midling OT. Fatigue properties of as-welded AA6005 and AA6082 aluminium alloys in T1 and T5 temper condition. Materials Park, OH: ASM International; 1996.

[24] Haagensen PJ, Midlin OT, Rane M. Fatigue performance of friction stir butt welds in a6000 series aluminum alloy. Trans Eng Sci 1995;8:589−98. Available from: https://doi.org/10.2495/SURF950271.

[25] Ericsson M, Sandström R. Influence of welding speed on the fatigue of friction stir welds, and comparison with MIG and TIG. Int J Fatigue 2003;. Available from: https://doi.org/10.1016/S0142-1123(03)00059-8.

[26] Vigh LG, Okura I. Fatigue behaviour of friction stir welded aluminium bridge deck segment. Mater Des 2013;44:119−27. Available from: https://doi.org/10.1016/j.matdes.2012.08.007.

[27] Moreira PMGP, de Figueiredo MAV, de Castro PMST. Fatigue behaviour of FSW and MIG weldments for two aluminium alloys. Theor Appl Fract Mech 2007;48:169−77. Available from: https://doi.org/10.1016/j.tafmec.2007.06.001.

[28] Di S, Yang X, Fang D, Luan G. The influence of zigzag-curve defect on the fatigue properties of friction stir welds in 7075-T6 Al alloy. Mater Chem Phys 2007;. Available from: https://doi.org/10.1016/j.matchemphys.2007.01.023.

[29] Uematsu Y, Tokaji K, Fujiwara K, Tozaki Y, Shibata H. Fatigue behaviour of cast magnesium alloy AZ91 microstructurally modified by friction stir processing. Fatigue Fract Eng Mater Struct 2009;32:541−51. Available from: https://doi.org/10.1111/j.1460-2695.2009.01358.x.

[30] Padmanaban G, Balasubramanian V. Fatigue performance of pulsed current gas tungsten arc, friction stir and laser beam welded AZ31B magnesium alloy joints. Mater Des 2010;31:3724−32. Available from: https://doi.org/10.1016/j.matdes.2010.03.013.

[31] Polezhayeva H, Toumpis AI, Galloway AM, Molter L, Ahmad B, Fitzpatrick ME. Fatigue performance of friction stir welded marine grade steel. Int J Fatigue 2015;81:162−70. Available from: https://doi.org/10.1016/j.ijfatigue.2015.08.003.

[32] McPherson NA, Galloway AM, Cater SR, Hambling SJ. Friction stir welding of thin DH36 steel plate. Sci Technol Weld Join 2013;18:441−50. Available from: https://doi.org/10.1179/1362171813Y.0000000122.

[33] Lakshminarayanan AK, Balasubramanian V. Assessment of fatigue life and crack growth resistance of friction stir welded AISI 409M ferritic stainless steel joints. Mater Sci Eng A 2012;539:143–53. Available from: https://doi.org/10.1016/j.msea.2012.01.071.

[34] Chabok A, Dehghani K, Jazani MA. Comparing the fatigue and corrosion behavior of nanograin and coarse-grain if steels. Acta Metall Sin (English Lett) 2015;28:295–301. Available from: https://doi.org/10.1007/s40195-014-0196-2.

Influence of Welding Parameters on Fatigue Behavior

3.1 INTRODUCTION

In Chapter 2, Fatigue Behavior in Friction Stir Welds, we discussed how the macro features and the associated microstructure of common friction stir welding (FSW) joint configurations can have a profound influence on the fatigue behavior. Of the chief factors that influence fatigue behavior, macro features like interfacial hooks, effective sheet thickness, weld bond width, and microstructural features (e.g., texture, grain size, grain orientation, grain boundary enrichment) play a crucial role in controlling the fracture behavior and ultimately the fatigue performance of welded joints. As one would expect, the macrostructure and microstructure features are highly sensitive to the FSW process. For example, a small change in tool plunge depth or tool tilt angle during the welding process can have a profound impact on the fracture and fatigue performance of a given FSW joint. Similarly, tool rotation rate, tool shoulder design, tool pin geometry, and material stacking configuration are all crucial parameters to obtain structural joints with superior fatigue performance. Interestingly, macro features in friction stir linear weld (FSLW) and microstructure in friction stir butt weld (FSBW) play a significant role in controlling the fatigue behavior of a given joint.

In this chapter, readers are introduced to the importance of identifying welding process parameters, and how they relate to fatigue behavior in a FSW joint. Crucial welding process conditions are identified in the early part of the design, and hence it is vital that engineers and researchers develop a comprehensive understanding on how fatigue performance in FSW is affected by the welding process parameters and conditions.

Once materials are identified during the initial stage of the design process, engineers determine the specific welding and material configuration that may effectively and efficiently produce FSW joints with

Fatigue in Friction Stir Welding. DOI: https://doi.org/10.1016/B978-0-12-816131-9.00003-9
© 2019 Elsevier Inc. All rights reserved.

superior strength and fatigue performance. The methodology involves developing a process window, which includes possible welding process variables and eventually identifying the optimum process parameters that may produce robust FSW joints. In FSW, the process parameters include several variables, and they can differ between a spot and linear weld and between overlap and butt weld configuration. Fig. 3.1 shows an overview of the important process variables that have been observed to impact the static and fatigue performance of an FSW joint. Thus, it is important that these variables are identified early in the design process based on the specific FSW type and joint configuration.

As widely known, FSW is a solid state joining process in which metallurgical bonding is established by plastic deformation of the material due to the downward forging force and the frictional heat generated between the weld tool shoulder and material surface. Regardless of the FSW type and joint configuration, the weld tool is the most significant part of an FSW process. Along with the tool shoulder and tool pin, the weld tool plays a dominant role in

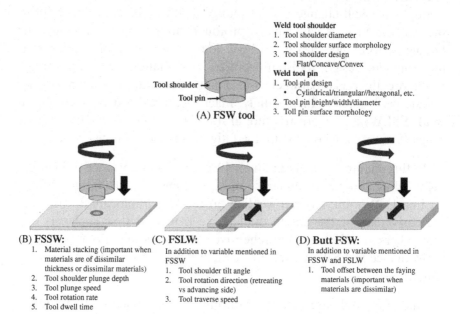

Figure 3.1 Illustration of various friction stir welding (FSW) variables associated with (A) a typical FSW tool, (B) friction stir spot weld (FSSW), (C) lap-shear friction stir linear weld (FSLW), and (D) Butt-friction stir weld (FSBW).

producing a mechanically and metallurgical strong weld. In addition to the weld tool, welding process conditions or parameters also play a key role in establishing a sound FSW joint. Nevertheless, a small change in any of these variables can have a significant impact on the mechanical performance, including the fatigue behavior of a given FSW joint. Therefore it is essential to develop a good understanding of the relationship between the process variables and its impact on the fatigue performance.

3.2 WELD TOOL DESIGN

In the FSW process, the weld tool plays a crucial role in producing not only the width but also the profile of the bond area in the weld. In its simplest form, an FSW tool consists of two important physical entities, the tool shoulder and the tool pin, also referred to as the probe pin. The tool shoulder and tool pin have different influences on specific aspects of an FSW joint. With increased interest and widespread use of FSW, new innovative weld tools are continuously being designed, where some use a consumable tool pin and some are designed with a pin-less weld tool. These FSW tools are designed for specific applications and materials. Additional innovations like retrievable pins have been designed to enable the material to fill in the keyhole or exit-hole feature that is made when the weld tool is removed during the termination of FSW. Generally, an FSW tool made of high strength tool steel has sufficient tool life when used in FSW of soft alloys such as magnesium and aluminum. There are also weld tools made of titanium or tungsten carbides [1] and Ferro-Titanit [2] for FSW of hard alloys of titanium and steel.

Of the several variables that can be altered in a weld tool, the tool shoulder and tool pin profile have a strong influence in controlling the macro and metallurgical features in an FSW joint. In fact, numerous studies have been performed on studying the effect of varying tool shoulder and pin profile on the static strength of the FSW joints. The tool shoulder with its relatively large surface area comes in direct contact with the material surface during FSW and thus generates enormous frictional heat. The downward forging force applied during FSW is also transferred to the weldment through the weld tool shoulder. Badarinarayan et al. [3] studied the effect of three different weld tool shoulder profiles on the static strength in friction stir spot weld

(FSSW) joints. The weld tools studied by Badarinarayan et al. are presented in Fig. 3.2. They observed that a weld tool with a concave tool shoulder profile produced FSSW with significantly higher effective sheet thickness, which drastically improved the weld strength compared to FSSW produced with a convex tool shoulder profile. During FSW, the concave profile of the tool shoulder assists in drawing the plastically deformed material from the stir zone upward toward the tool shoulder, thereby considerably improving the bulk material movement toward the free surface. This flow of bulk material upward significantly reduced the hook height and simultaneously increased the effective sheet thickness during FSSW. In an early study on fatigue and fracture behavior of FSSW produced using two unique tool shoulder profiles, Lin et al. [4,5] showed that an FSSW produced with a concave tool shoulder exhibited slightly better fatigue performance at higher cyclic loads when compared to welds produced with a flat shoulder. The effect of a smaller interfacial hook and large effective sheet thickness in improving the fatigue performance of an FSW joint has also been demonstrated in other studies [6–10]. In addition to tool shoulder profile, tool shoulder diameter can also influence the amount of frictional heat generated. With the increase in tool shoulder diameter, the frictional heat generated increases due to increased availability of frictional surface and thereby considerably improving the plastic deformation of the material. The shoulder modification has been observed to improve the static strength in FSW joints [11]. In addition to tool shoulder diameter, surface morphology of the tool shoulder is also of interest. de Giorgi et al. [12] studied the fatigue performance of FSW AA6082 FSBW joints produced using three different tool shoulder surface morphologies; a shoulder with a scroll, a shoulder with a shallow cavity, and a flat shoulder. He observed that the weld tool

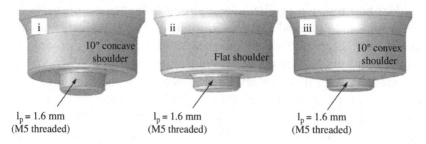

Figure 3.2 Schematic illustrating the three shoulder profiles used to produce friction stir spot weld in aluminum 5754-O sheets [3].

shoulder with scroll morphology produced a rough surface along the weld line from where the dominant fatigue cracks initiated, lowering the fatigue life compared to welds produced with a shallow cavity and flat shoulder weld tools. Beyond the work of Giorgi, the majority of the research to date on the influence of tool shoulder geometry, including the shoulder profile, surface morphology, and geometry, is limited to static strength. Generally, most researchers have preferred to use a concave tool shoulder with no surface morphology in order to produce FSW joints with good fatigue performance [12−16].

In addition to the tool shoulder, the tool pin also has an important role in material mixing and material flow during FSW. While the tool shoulder contributes to the downward forging force and the generation of frictional heat needed for the plasticization of the material, the tool pin penetrates the material and contributes to the material mixing in the stir zone. One of the most common pin tool designs is the simple cylindrical pin profile. However, the continued quest for stronger FSW joints has led researchers and engineers to design new tool pin profiles. Some of the initial research on the influence of tool pin geometry in overlap spot welds included using a triangular pin geometry. In the studies by Badarinaryan et al. [3,17], FSSW overlap joints fabricated with a triangular pin profile (Fig. 3.3B) exhibited superior static strength when compared to welds produced with a cylindrical tool pin (Fig. 3.3A). This is in part because the triangular pin produced finer grains in the stir zone and greatly increased the material deformation. Ultimately, this resulted in larger effective sheet thickness, which enhanced the fatigue crack growth response of the FSSW [13].

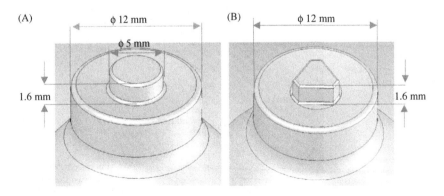

Figure 3.3 Schematic illustrating of friction stir welding (FSW) tool geometries (A) cylindrical pin profile and (B) triangular pin profile [17].

In another study on FSSW, Yin et al. [18] produced several FSSW joints by use of three different tool designs and varying tool rotation rates. Interestingly, each of the weld tool designs produced unique interfacial hooks, even when comparing those created at identical tool rotation rates. Their results are depicted in Fig. 3.4. Recalling from Chapter 2, Fatigue Behavior in Friction Stir Welds, interfacial hooks play a significant role in controlling the fatigue properties in FSSW where a smaller interfacial hook is generally observed to improve the overall fatigue performance in FSSW joints [13]. In addition to tool pin profile, the surface or features of the tool pin also influence the macro features. Several studies have indicated that welds produced with a threaded tool pin profile generally exhibited higher fatigue strength compared to welds produced with an unthreaded tool pin profile [14,19,20]. The improvement in fatigue performance of the welds produced with a threaded tool pin profile is attributed to superior material mixing and material flow in the stir zone and producing

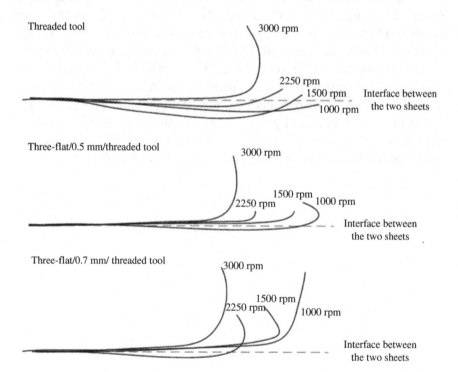

Figure 3.4 Schematic representation showing interfacial hook geometry produced under various tool rotation rates in each of the weld tool pin designs [18]. Reprinted with permission from Taylor and Francis.

favorable microstructure in the weldment compared to welds produced with an unthreaded tool pins [14,19−22].

Beyond the surface morphology and tool pin design, research on the length and width of the tool pin has also been an area of interest. With the goal of achieving sound welds, it is important to maintain a weld tool with optimum pin length, particularly in production of overlap FSW joints. The tool pin should have sufficient length to enable the pin to penetrate the top sheet completely and reach the surface of the bottom sheet/material in overlap joints before plunging the weld tool. If the tool pin does not completely penetrate the top sheet and reach the surface of the bottom underlying sheet/material during the welding process, it will result in insufficient material mixing and ultimately lead to inferior quality welds. Weld tools designed with smaller or inadequate tool pins will need a deeper tool plunge depth during FSSW. Though sufficient tool plunge is a requirement during FSW, excess tool plunge can be detrimental to weld performance. The impact of tool plunge depth is discussed later in this chapter.

3.3 EFFECT OF KEYHOLE FEATURE

The keyhole or exit-hole feature is a distinctive and significant characteristic of any FSW joint. The keyhole is formed inevitably as a result of extraction of the weld tool during the termination of the FSW process. In the case of the FSSW, these keyholes become a permanent physical entity of the weld. Under fatigue loading, the keyhole may nucleate fatigue cracks and lead to fracture of the welded joint [23]. In recent years, researchers have devised innovative methods to overcome and address the problems associated with the keyhole feature. Note that in FSBW and overlap FSLW, the keyhole may be removed from the weldment by terminating the weld on a tab, which can be machined off post-weld. However, in some structural joints and particularly in the case of the FSSW, cutting out the keyhole is not feasible. As such, researchers have designed semi-consumable tools which can refill the keyhole [24,25] and a pin-less weld tool [26−29] to address this issue. Furthermore, re-filling processes that use a retractable pin tool have been developed to push material back into the keyhole during the final stage of FSSW. The effect on the re-filling technique on fatigue performance in FSSW is inconclusive [30]. Fig. 3.5 shows the representative cross-sections of FSSW joints produced with and

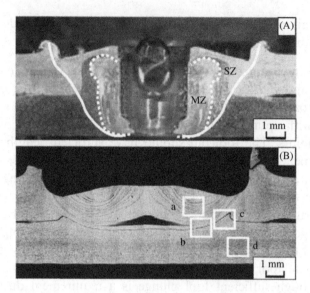

Figure 3.5 Cross-section of a friction stir spot weld produced using (A) a weld tool with a tool/probe pin and (B) tool/probe pin-free weld tool highlighting the (a) stir zone region, (b) boundary between the lower and upper sheet region, (c) boundary region of the sheets at outer circumference of the shoulder, and (d) parent material region [29].

without a tool/probe pin [29]. While the specimen in Fig. 3.5A has the distinctive keyhole, the specimen in Fig. 3.5B is produced without the keyhole.

Although there are instances of fatigue crack initiating at the keyhole in butt-welded joints during cyclic loading, the prevailing role of the keyhole and its influence on fatigue life in lap-joints is not completely understood. In a recent study on FSLW aluminum to magnesium alloys, Rao et al. [31] observed that the keyhole had a minimal role in nucleating fatigue cracks. It was observed that although the test specimens with the keyhole exhibited lower fatigue life, the decrease in the overall weld length due to presence of the keyhole contributed to lower fatigue life. Representative results from this work are provided in Fig. 3.6 [31]. Further work is needed to determine the effect of the keyhole in FSLW joints of other materials.

Given the role that tooling plays in the FSW process, the influence that the pin and shoulder have in producing welds with good mechanical and metallurgical properties is not all that surprising. In combination, the tool shoulder and tool pin must provide adequate frictional heat and material flow, in turn directly influencing the geometry of

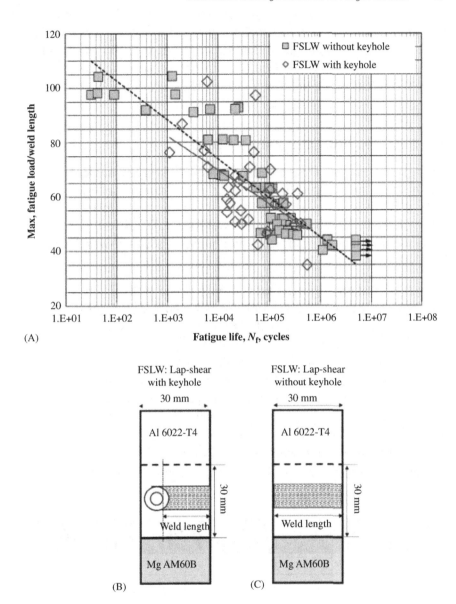

(A)

(B)   (C)

Figure 3.6 (A) Fatigue life versus ratio of maximum fatigue load to weld length. (B) Representation of the weld length in welds with keyhole, and (C) in welds without keyhole [31].

macro features such as the interfacial hooks, defects, weld surface roughness in addition to residual stresses, and microstructure of the FSW. The macro features, residual stresses, and microstructure subsequently influence the fracture and fatigue behavior of the FSW joints. Therefore, FSW tool design and optimization can be considered the first and most basic step in development of the FSW process window. Once the FSW tool design is identified for a given FSW joint and the material combination, the next step in the process window development is to identify the optimal welding process parameters.

3.4 WELDING PROCESS PARAMETERS

FSW process parameters consist of several different variables that vary for different joint types. For a better understanding of the different FSW joint types, we will divide the processing parameters that are common and specific to respective weld types. Table 3.1 provides an overview of the welding process parameters associated with a particular FSW joint.

It is important to note that although some welding process parameters such as tool rotation rate and tool plunge depth can share similar values across the different FSW and joint types, they are not always interchangeable. For example, an optimum tool rotation rate identified for an FSSW in a particular material may not be optimum in producing FSLW or FSBW with the same material combination. Therefore the common welding process parameters or variable should not be confused with the defined, numerical welding parameters that are optimized or identified for a given FSW and joint type. In addition to welding process parameters specified in Table 3.1, the material stacking configuration is particularly important when using FSW to join dissimilar materials.

Table 3.1 Critical FSW Process Parameters That Are Common and Specific to FSW and Joint Types			
Welding Process Parameters	FSBW	FSLW	FSSW
Tool rotation rate (rpm)	X	X	X
Tool shoulder plunge depth (mm)	X	X	X
Tool shoulder plunge rate (mm/min)	X	–	X
Tool dwell time (s)	X	X	X
Tool traverse speed (mm/min)	X	X	–
Tool tilt angle (in degrees)	X	X	–
Tool rotation direction (with respect to traverse direction)	X	X	–
Tool offset from weld line	X	–	–
X, critical parameter; –, not applicable.			

The established understanding in FSW is that the majority of the frictional heat is produced by the interaction of tool shoulder with the material surface. In addition to tool shoulder geometry, tool rotation rate, dwell time in FSSW, and tool traverse speed in FSLW and FSBW influence the amount of frictional heat generated in a FSW. Thus it makes sense that with an increase in tool rotation rate, the amount of frictional heat generated increases. An important fundamental understanding to note is that frictional heat is necessary for adequate plastic deformation, material flow, and mixing during the FSW process. Low frictional heat may result in poor material flow and plastic deformation resulting in the formation of micro voids in the weldment. These micro voids reduce the static strength of the weld joint or nucleate fatigue cracks [32]. Similarly, a very high frictional heat will facilitate excess material flow during welding and consequently lead to material being expelled as "flash." The excess flow may result in the formation of cavities in the weldment in addition to producing large heat affected zone (HAZ). The effect of tool rotation on frictional heat generated is coupled with the tool dwell time (in FSSW) and tool traverse speed (in FSLW and FSBW). That is, a high tool rotation rate and low traverse speed or longer dwell time would substantially increase the frictional heat in the weldment and may not produce anticipated macrostructural or microstructural features. Accordingly, a low tool rotation rate coupled with a high tool traverse speed or shorter dwell time would generate inadequate frictional heat resulting in poor quality FSW joints. James et al. [33] reported that by increasing the tool traverse speed, the endurance limit of the FSBW 5XXX aluminum alloys reduced compared to joints produced at lower tool transverse speeds. An example of how the tool rotation rate and the tool traverse speed affect the macro features in an FSLW is presented in Fig. 3.7 [32]. The presence of voids or cavities in the weldment or stir zone substantially reduces the fatigue life in an FSW joint [32,33]. As we know, fatigue cracks tend to nucleate from preexisting defects; therefore it is essential that an optimum tool rotation rate and tool traverse speed/dwell time be identified for a given weld and joint type [20,22,32−35].

In a study by Naik et al. [36] on FSLW of AZ31 magnesium alloy, a change in tool rotation rate and tool traverse speed affected the geometry of the interfacial hooks, as shown in Fig. 3.8. Fatigue tests of the samples produced using 1000 rpm rotational speed and 20 mm/s

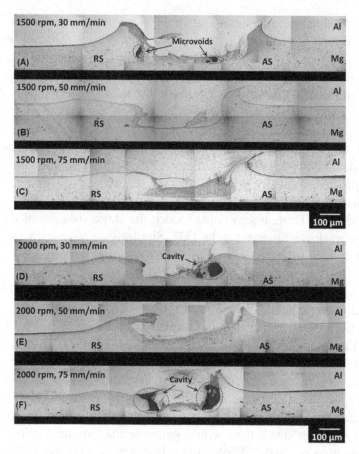

Figure 3.7 Cross-section of lap-shear FSLW aluminum 6022 to magnesium AM60 alloy produced under varying tool rotation rates and tool traverse speeds: (A) 1500 rpm and 30 mm/min traverse speed, (B) 1500 rpm and 50 mm/min traverse speed, (C) 1500 rpm and 75 mm/min traverse speed, (D) 2000 rpm and 30 mm/min traverse speed, (E) 2000 rpm and 50 mm/min traverse speed and (F) 2000 rpm and 75 mm/min traverse speed [32].

tool traverse speed exhibited superior fatigue performance compared to samples produced at 1500 rpm rotational speeds with 10 and 20 mm/s tool traverse speeds, as indicated in the plot in Fig. 3.9 [36]. The increase in fatigue performance of the FSLW was attributed to the favorable interfacial hook geometry in the specimens produced at 1000 rpm rotational speed and 20 mm/s tool traverse speed. This is in contrast with the welds produced at the higher rpm which produced sharp tipped interfacial hooks that nucleated dominant fatigue cracks leading to lower number of cycles to failure. The geometry of the hook, specifically on the advancing side of the weld, substantially reduced the effective sheet thickness in welds produced at 1500 rpm.

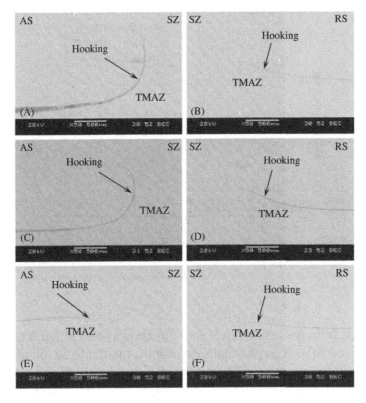

Figure 3.8 Interfacial hooks observed on the advancing (left) and retreating (right) sides at varying tool rotational rates and welding speeds, namely (A) to (B) 1500 rpm and 10 mm/s, (C) to (D) 1500 rpm and 20 mm/s, and (E) to (F) 1000 rpm and 20 mm/s [36]. Reprinted with permission from Springer Nature.

This indicates that at a higher tool rotation rate, a greater amount of frictional heat is produced facilitating material flow and reducing the overall effective sheet thickness on the advancing side of the weld. Tool rotation rate also impacts the interfacial hook geometry in FSSW. Referring to Fig. 3.4, it is evident that increasing the tool rotation rate irrespective of tool pin geometry can produce large interfacial hooks. The large interfacial hooks, in turn, adversely affect the static strength and corresponding fatigue life in FSSW joints.

Tool shoulder plunge depth is the amount of depth a tool shoulder is plunged into the material surface before the weld tool begins to dwell or traverse. It is a numerical quantity and is the distance measured from the original material surface to the surface of the keyhole or exit-hole in an FSW joint. During initial process development, the plunge depth is determined based on the thickness of the materials that

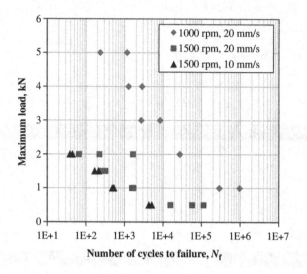

Figure 3.9 Fatigue life curves of the friction stir linear weld produced using different process parameters [36]. Reprinted with permission from Springer Nature.

is welded. With the length of the tool pin known, the plunge depth must be sufficient enough to plastically deform the material and initiate material flow in the stir zone of the weldment. Insufficient plunge depth will cause partial bonding between the overlapping sheets or butt joint generally resulting in poor quality welds. There are no reliable studies on the impact of plunge depth on fatigue performance in FSW joints. However, based on the effect of plunge depth on the macro features, we provide the following observation: in FSSW and FSLW, plunge depth is significant in controlling the height of the interfacial hook and effective sheet thickness which impacts the fatigue life. A deeper plunge depth will drive the plastically deformed material upward toward the tool shoulder and ultimately be expelled as "flash." Generally, increasing the plunge depth produces smaller interfacial hooks and larger bond width. Both of these effects significantly improve the static strength of the FSW joints [37−39]. Beyond a certain plunge depth, top sheet thinning occurs reducing the effective sheet thickness [37,38], which is detrimental to the fatigue performance of FSLW joints. In certain FSW joints where the weld tools are designed without a tool pin, a deeper plunge depth improved the static strength of the weld joint [27,29]. It is important to note that the macro features, specifically the interfacial hooks in a pin-less FSW lap joint, are quite different from FSW lap-joints produced with a tool pin.

Another aspect of FSW parameters to consider is the plunge rate. During FSW, the tool shoulder is plunged onto the material surface at a prescribed rate. This welding process parameter is critical for FSSW and the transient start-up section of welds produced by FSLW and FSBW. One study found that the effect of tool plunge rate was only observed when the tool rotation rate for a given FSSW was not optimized [39]. In one study on the FSSW of dissimilar metals (magnesium to aluminum), a slow plunge rate resulted in higher temperatures in the weldment, and with a gradual increase in plunge depth rate, the temperature decreased even when the tool rotation rate and dwell time were kept constant [40]. As the tool plunges slowly into the top sheet, the tool pin generates localized friction heat even before the tool shoulder comes in contact with the sheet material surface. Additional frictional heat is generated on further plunging of the tool shoulder until the required plunge depth is achieved. When the tool plunge rate is faster, the tool pin and tool shoulder have less time to interact with the material before reaching the desired plunge depth. This results in lower temperatures within the weldment. Based on the materials being welded and their metallurgical sensitivity to temperature, tool plunge rate does impact the microstructure features, which can ultimately affect the fatigue performance of a given FSSW. Therefore it is important to know the material sensitivity to temperature before identifying the plunge rate.

The next parameter to consider is the tool tilt angle. In FSLW and FSBW, after plunging the material surface to a predetermined depth (plunge depth), the weld tool is made to traverse a certain distance along the weld line to obtain an FSW in overlap or butt weld configuration. Generally, in most butt and overlap FSW processing, the tool shoulder is not held parallel to the material surface. In fact, the tool is typically tilted to a small angle θ ($0° < \theta > 5°$) as illustrated in Fig. 3.10A. This small but crucial tilt angle of the tool ensures the flow of plasticized materials from the front of the tool pin to the back of the tool pin, thereby filling the cavity created on the advancing side of the weld [41]. Increasing the tool tilt angle provides sufficient space for the plasticized material to be retrieved back into the stir zone. As an example, for FSW of dissimilar materials, increasing the tool tilt angle resulted in improved material mixture. Unfortunately, the increased tilt angle also increased the thickness of intermetallic compounds (IMCs) formed in the weld [42,43], which drastically decreased

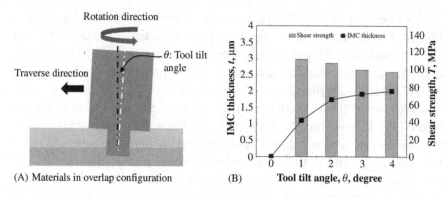

Figure 3.10 (A) Pictorial illustration of the tool tilt angle during the friction stir welding (FSW) of materials in overlap configuration (friction stir linear weld, FSLW), and (B) relationship between the shear strength, intermetallic thickness, and tool tilt angle in FSLW aluminum to steel alloys. Data replotted from Kimapong K, Watanabe T. Effect of welding process parameters on mechanical property of FSW lap joint between aluminum alloy and steel. Mater Trans 2005;46:2211−7.

the static strength of the dissimilar FSLW joint (Fig. 3.10B). In FSW of pure titanium, Seighalani et al. [44] observed that FSW butt joints produced at 1° tool tilt angle exhibited better weld strength compared to welds produced at 0° tool tilt angle. The work of Elyasi et al. [45] showed that FSW butt joints produced with a 2° tool tilt angle exhibited better fatigue performance when compared to welds produced at 1° and 3° tool tilt angles. In fact, this research found that too large of a tool tilt angle resulted in ejection of the plasticized material from the advancing side as "flash." This resulted in insufficient material being retrieved back into the stir zone and thereby introducing defects in the weld nugget. Furthermore, a large tilt angle can also reduce the effective sheet thickness of the material on the advancing and retreating sides of the FSW [46].

Once the weld tool is plunged and tilted to the required angle, the tool is then traversed along the weld line for a predetermined distance in order to obtain an FSLW or an FSBW joint. One important welding process parameter that must be determined is the tool rotational direction. The tool rotation may either be clockwise or counterclockwise with respect to the tool traverse direction. While the selection of tool rotation direction might seem trivial, the tool rotational direction may have a profound impact on the fatigue behavior in FSLW and dissimilar FSBW joints. To better understand the impact of tool rotation direction, consider the schematics in Fig. 3.11A and D illustrating the FSW of materials in butt joint configuration (as seen from above).

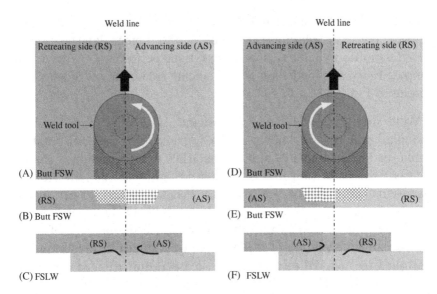

Figure 3.11 Pictorial illustration of tool rotation with respect to tool traverse direction for the (A) counterclockwise and (D) clockwise rotation. Schematic of the cross-section indicating the different microstructure formed in friction stir butt weld due to change in direction for (B) counterclockwise and (E) clockwise tool rotation. Schematic of macro features in overlap friction stir linear weld joints formed in overlap friction stir welding (FSW) joints for (C) counterclockwise and (F) clockwise tool rotation.

The bold black arrows in the two figures indicate tool traverse direction and the yellow arrows indicate the tool rotation direction. In Fig. 3.11A, the tool is rotating counterclockwise, and in Fig. 3.11D the tool is rotating clockwise with respect to the tool traverse direction. This change in weld tool rotation will impact the formation of macro features, specifically the advancing side hook and retreating side cold lap feature in overlap FSLW and the microstructure in FSBW. This is schematically illustrated in Fig. 3.11B and C for welds produced with counterclockwise tool rotation and in Fig. 3.11E and F for welds produced with clockwise tool rotation. The formation of an advancing side hook and the retreating side cold lap features in FSLW along with the microstructure produced in the FSBW are explained in Chapter 2, Fatigue Behavior in Friction Stir Welds. Given that the fatigue cracks tend to nucleate from one of these macro features (technically defects) in FSLW joints, it is important that we produce welds with optimum macro features. Research on tool rotation direction has shown that overlap FSLW joints which produced the macro features illustrated in Fig. 3.11C exhibited higher static strength when compared to the FSLW joints that produced the macro features shown in Fig. 3.11F [9,38,47,48]. Furthermore, Moraes et al. [9] observed that weld coupons of FSLW magnesium AZ31 that were loaded on the retreating side cold lap feature exhibited a 50% higher lap-shear static strength and up to a 300% increase in fatigue performance (Fig. 3.12B) compared to welds loaded on the advancing side hook, as indicated by the plot in Fig. 3.12A.

Weld strength and fatigue performance of FSBW joints made of dissimilar materials is known to be impacted by the tool rotation. Unlike in FSBW joints of similar materials, when welding dissimilar materials the formation of IMCs must be mitigated in order to produce welds with superior weld strength. IMCs are brittle in nature, and thus the formation of IMCs in the weld nugget is likely to substantially reduce the static strength and ultimately reduce the number of cycles to failure. The formation of IMCs in the weld nugget is predominantly controlled by material mixing and the temperature attained during the welding process. Several studies have observed that dissimilar FSBW joints of aluminum and magnesium alloys have an increased static strength when the aluminum alloy was placed on the advancing side [49–52]. The reduction in strength when the magnesium alloy was placed on the advancing side was due to the increased heat input during welding. The increased heat plasticized the material more severely, but also resulted in increased IMCs formation in the weld nugget. If the material placed on the advancing side is harder and/or of higher strength compared to material on the retreating side, the total frictional heat input to the system is greater when compared to a softer material placed on the advancing side. Therefore it is important to

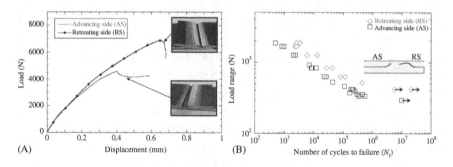

Figure 3.12 (A) Representative load versus displacement curves of friction stir linear weld (FSLW) lap-shear tests of the retreating side (RS) and advancing side (AS) orientated coupons and (B) experimental results of load range versus the number of cycles to failure of the FSLW lap-shear coupons tested at a load ratio R = 0.1 [9].

identify the material stacking/placement and the tool rotation direction when creating FSBW joints of dissimilar materials.

In addition to tool rotation direction, the use of tool pin offset during welding can also impact the formation of IMCs. Yamamoto et al. [53] studied the influence of tool pin offset during FSBW joint creation of aluminum to magnesium alloys. In their study, the aluminum alloy was placed on the advancing side and the tool pin was slightly offset into the aluminum to produce the FSBW joint. As the offset distance increased from the weld line, the thickness of the IMCs in the weld nugget also increased considerably, as shown in Fig. 3.13A. The researchers showed that the FSBW joint with a thicker formation of IMCs exhibited lower tensile strength compared to welds with the thinner layer of IMCs (Fig. 3.13B). Similar observations with respect to material placement and tool offset were also observed in FSBW joints of aluminum and copper [54−56]. Thus we can conclude that material position/stacking, tool rotation direction, and tool pin offset during the creation of FSBW joints of dissimilar welds can be used to tailor the mechanical strength of these joints. While the authors are not aware of any published studies on the effect of tool rotation direction and tool pin offset on fatigue behavior in dissimilar metals, the authors conclude that the greater the volume of IMCs in the weld nugget the lower the fatigue life would be due to the presence of brittle IMCs [57]. Therefore it is vital that one chooses the appropriate material position/stacking, tool rotation direction, and tool pin offset during FSBW of dissimilar materials.

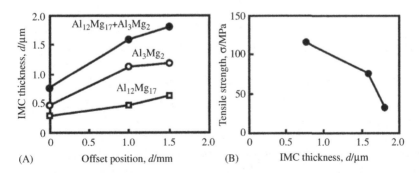

Figure 3.13 (A) Thickness of intermetallic compounds (IMCs) versus the tool offset distance, and (B) thickness of the IMCs versus the tensile strength in friction stir butt weld aluminum to magnesium alloys [53]. Reprinted with permission from the Japan Institute of Metals.

So far the readers have been introduced to various critical FSW process parameters that influence the static as well as fatigue and fracture behavior of many FSW joints. It is important to note that most of the literature available on the fatigue of FSW joints are based on observations made in FSW specimens that were produced with optimum welding process parameters. In most cases, the criterion for selecting the optimum FSW process parameters is superior static strength. This is also a common methodology in most other common welding processes like resistance spot welding, ultrasonic, and laser welding. However, it is important to understand that in FSW, selection of optimal process parameters by use of static strength may not always provide the best fatigue behavior. This is due to the fact that there are several features that can impact the fatigue life. We may be able to produce an FSW joint with no evidence of visual defects, but the microstructure will likely play a critical role during fatigue crack initiation and propagation. Additionally, the FSW defects (including interfacial hooks, voids/cavities, keyhole, and weld surface roughness) may prematurely nucleate the dominant fatigue crack, yet may not necessarily affect static strength. One must also consider the microstructure features and residual stresses in the weld joint, which can ultimately impact the rate at which the dominant fatigue crack propagates through the joint. Hence, the laboratory generated monotonic test results should be considered as a baseline for choosing the optimum welding process parameters, but a robust fatigue testing plan of the optimally produced FSW joints is recommended.

3.5 RESIDUAL STRESSES

In the previous section, we discussed the various aspects of developing a process window and how the individual welding parameters can affect the fatigue performance of an FSW joint. Many structural joints can fracture without the application of an external load. In such cases, residual stresses are usually blamed for failure. Residual stresses can be introduced in the material during manufacturing processes such as casting, forging, welding, heat treatment, rolling, bending, and surface treatment. In addition, residual stresses can develop in materials during assembly and under acceptable operating conditions. Furthermore, the development of thermal gradients and volumetric changes arising during solidification, or from differences in coefficient of thermal expansion in dissimilar materials, may also introduce residual stresses in the material. We should ask—"Are residual stresses bad for a structural material?" The answer depends on the type of residual stress manifested in a material. Brittle materials can be toughened by introducing compressive residual stress. Tensile residual stresses may cause early fracture, as they aid in crack tip opening during fatigue loading. The residual stresses manifested during FSW are primarily thermally induced and can be introduced as a result of any of the processing variables we discussed earlier in this chapter. Additionally, external factors such as the clamping forces that hold the materials in place during welding and the thermal conductivity of the base plate on which the materials are secured during welding can also influence the development of residual stresses in FSW joints [58]. Since residual stresses can be difficult to accurately measure and may accelerate fatigue crack initiation and propagation, it is important that we develop a basic understanding of what factors contribute to residual stresses in FSW joints and how they influence the corresponding fatigue performance.

Compared to fusion welding processes, the temperatures attained during the FSW process are below the base material solidus, thereby helping to maintain dimensional stability after welding. Nevertheless, residual stresses can still develop during FSW. Richter-Tummer et al. [58] observed that a well selected clamping force produced an FSW joint with lower distortion and uniform distribution of residual stress compared to FSW joints produced with a low clamping force (Fig. 3.14). The FSW joints produced with a higher clamping force exhibited better mechanical performance compared to FSW joints produced with a low clamping force.

Beyond the clamping fixture design and forces, the FSW process parameters may also influence the development of residual stresses in the FSW joints. Reynolds et al. [59] noted that in FSW 304L stainless steel welded at 300 and 500 rpm tool rotation rate, the magnitude of the longitudinal and traverse residual stresses were similar even though they exhibited a slightly different microstructure. Additionally, the transverse residual stress changed from a tensile residual stress near the crown to a compressive residual stress at the root of the stir zone. This in turn introduced distortion in the weld joint. Thus a common observation in residual stress measurements in FSW is that typically the stir zone, which experiences severe plastic deformation and frictional heat during welding, generally exhibits tensile residual stresses while the

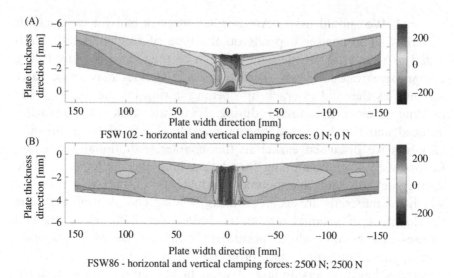

Figure 3.14 Comparison of residual stress distribution in an friction stir welding (FSW) joint fabricated with no (A) clamping forces and (B) fabricated with clamping forces. Residual stresses in [MPa] [58].

base metal commonly exhibits a compressive residual stress [60−63]. Furthermore, while residual stresses resulting from FSW are generally relatively low compared to fusion-based welding techniques, the magnitude of the residual stresses in FSW can represent a significant fraction of the base material strength. In fact, in FSW of oxide dispersion strengthened steel, Brewer et al. [64] found that residual stresses in the thermo mechanically affected zone and HAZ were as high as 80% and 45% of the base materials strength, respectively.

During the initial welding process development stage, oftentimes the development of residual stresses is not considered in the FSW process window development. This is because quantifying residual stresses is a difficult and time-consuming task. With the advancement of FSW and its widespread application, many researchers are utilizing nondestructive methods to measure the residual stresses in the weld such as synchrotron [65,66]. James et al. [65] suggests considering residual stress measurement in laboratory specimens and extrapolating the data to larger structures which will help in designing and fine tuning the welding process parameters. As we discussed earlier in this chapter, welding process parameters can directly affect the amount of frictional heat generated during FSW. Specifically, tool rotation rate and tool traverse speed have been identified as the main contributors to residual stress development in the FSW joint [60,67,68]. As such, some techniques have been suggested to reduce the residual stresses in FSW. These techniques include (1) the use of external cooling during the FSW process to remove excess heat, (2) forced cooling of the base plate (anvil) to extract excess heat, and (3) post-weld shot peening [61,69]. The post-weld shot peen process is intended to introduce beneficial compressive residual stresses to offset the tensile residual stresses developed during welding. The compressive residual stresses in turn help to yield better fatigue crack growth resistance [70]. Further discussion of effect residual stress on fatigue crack growth rates in FSW is presented in Chapter 4, Fatigue Crack Growth in Friction Stir Welds.

3.6 STRENGTHENING MECHANISMS

Structural materials that are produced through conventional manufacturing methods like casting, forging, or machining generally undergo secondary process such as heat treatment, work hardening, or shot peening in order to improve the strength of the material. Post-strengthening mechanisms are used to mitigate the effects of residual stresses or to alter the microstructure of the material through grain refinement, introduction of dislocations, and precipitates. Strengthening mechanisms are not restricted to structural materials produced by conventional manufacturing processes but are also extensively used on weld and mechanical joints. During the FSW process, the base material undergoes severe plastic deformation under elevated temperatures, which produces microstructures in the weldment that are different from that of the base material. Microstructural changes include grain refinement and grain boundary enrichment, to name a few. In addition to microstructure, residual stresses are also introduced in the weldment as discussed in the previous section. Normally, most FSW joints are used in the as-welded condition, but secondary strengthening mechanisms can be performed to further improve the mechanical performance of the FSW joint. The two major strengthening mechanisms used in FSW joints are post-weld heat treatment and peening. During FSW, the temperatures reached during welding is much lower than the melting temperature of the base material but are still high enough to dissolve precipitates that strengthen the material [71]. Dissolution of strengthening precipitates particularly in 2XXX, 6XXX, and 7XXX series aluminum alloys results in material softening and will eventually lead to poor weld performance. To overcome this problem, post-weld heat treatments are performed on many of the

FSW joints. In aluminum 6XXX series Al-Mg-Si alloys, Mg_2Si is the major precipitate that strengthens the alloy. During the FSBW of aluminum 6XXX series alloys, it was found that FSW dissolves most of the precipitates resulting in poor tensile strength of the joints compared to the base material as seen in Fig. 3.15A and B [72]. The FSBW joints were subjected to post-weld heat treatment including solution treatment, which further dissolved most of the precipitates as observed in Fig. 3.15C. Solution treatment followed by aging formed a super saturated solid solution with precipitates reemerging at grain boundaries as

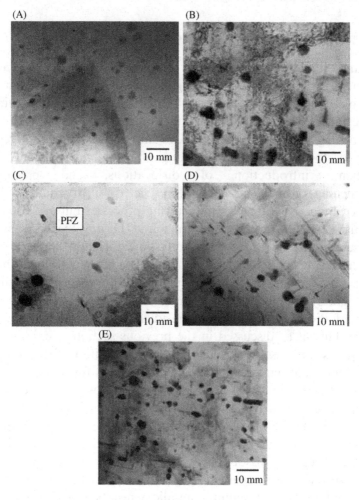

Figure 3.15 Distribution of strengthening precipitates in friction stir welding (FSW) zones. (A) Base metal, (B) As-welded (AW) joint, (C) solution-treated (ST) joint, (D) solution-treated and aged (STA) joint, and (E) artificially aged (AG) joint [72].

seen in Fig. 3.15D. Artificially aging the as-welded FSBW specimens produced fine precipitates uniformly distributed through the matrix as seen in Fig. 3.15E. The artificially aged FSBW specimens exhibited enhanced hardness and improved tensile strength of the joint compared to as welded and other heat-treated FSBW joints.

Similar observations were also made in other 7XXX series aluminum alloys where FSW dissolved most of the strengthening precipitates which later reappeared post-weld heat treatment to improve the mechanical performance of the FSBW joints compared to as-welded FSBW joints [73,74]. Although research on post-weld heat treatment of FSBW joints have shown to improve the weld strength, in some FSBW joints of aluminum, post-weld heat treatment rather decreased the weld strength [71,75–77]. In a study on effect of heat exposure on the fatigue performance of aluminum 7XXX series alloy, White et al. [77] observed that the FSBW specimens that were heated in air at 315 °C for 20 min exhibited fatigue life that was lower when compared to the as-welded FSBW specimens as seen in Fig. 3.16. The reduction in strength and ductility of the heat-exposed FSBW joints were attributed to over-aging and a partial annealing process that the joints underwent during the heating. The studies suggest that some FSW joints, particularly FSW aluminum alloys, are sensitive to heat damage.

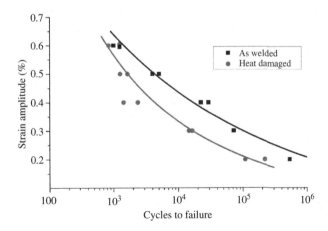

Figure 3.16 Experimental strain–life results for as-welded and heat-damaged AA7050 friction stir welding (FSW) [77]. Reprinted with permission from Springer Nature.

Strengthening mechanisms like shot peening and laser peening are mostly performed with a predetermined tension to mitigate the residual stresses in the FSW. In addition, it is important that at the designing and process development stage, engineers know how the FSW joints behave under operating temperatures where they can potentially undergo unintended post-weld heat treatment. This is critical in context that there are instances where an FSW joint will undergo some form of unforeseen heat treatment in-service or post-production. An example is when the automotive joints are produced in design and assembly stage referred to as body in white. Once the sheet materials of the body are welded and assembled, they are prepared for painting subsequently, in which the assembled body is baked (paint bake process) to temperatures close to 180°C for certain periods of time (approximately 30 min) and then cooled to room temperature. The actual paint bake temperature and exposure time, which are preparatory in nature varies between manufactures and models. Similarly, the FSW structures may be used in-service where they are exposed to temperatures which are enough to heat treat the weldment. Hence, it is crucial that design engineers know the sensitivity of the weldment to heat treatment and should be included in the very early stage of design.

3.7 SUMMARY

Developing an optimal FSW process for a given material and joint type is exhaustive with many process parameters that need to be individually analyzed during and after the welding process. While FSW is a very robust process and produces welds that exhibit high fatigue performance, the process can be seemingly sensitive to a miniscule variation in any one of the welding processes, and thus may have a significant impact on the mechanical properties including fatigue performance. Therefore we have made our best efforts to provide a brief discussion of the impact of various FSW process parameters on the fatigue performance in various FSW joints and weld types. Readers are highly encouraged to refer to the existing published work on FSW for more in-depth information and to remember that new studies on FSW are published nearly daily.

REFERENCES

[1] Edwards P, Ramulu M. Identification of process parameters for friction stir welding Ti−6Al−4V. J Eng Mater Technol 2010;132:031006. Available from: https://doi.org/10.1115/1.4001302.

[2] Pirondi A, Collini L. Analysis of crack propagation resistance of Al-Al2O3 particulate--reinforced composite friction stir welded butt joints. Int J Fatigue 2009;31:111—21. Available from: https://doi.org/10.1016/j.ijfatigue.2008.05.003.

[3] Badarinarayan H, Shi Y, Li X, Okamoto K. Effect of tool geometry on hook formation and static strength of friction stir spot welded aluminum 5754-O sheets. Int J Mach Tools Manuf 2009;49:814—23. Available from: https://doi.org/10.1016/j.ijmachtools.2009.06.001.

[4] Lin P, Pan J, Pan T. Failure modes and fatigue life estimations of spot friction welds in lap-shear specimens of aluminum 6111-T4 sheets. Part 1: welds made by a concave tool. Int J Fatigue 2008;30:74—89. Available from: https://doi.org/10.1016/j.ijfatigue.2007.02.016.

[5] Lin P, Pan J, Pan T. Failure modes and fatigue life estimations of spot friction welds in lap-shear specimens of aluminum 6111-T4 sheets. Part 2: welds made by a flat tool. Int J Fatigue 2008;30:90—105. Available from: https://doi.org/10.1016/j.ijfatigue.2007.02.017.

[6] Rao HM, Jordon JB. Effect of weld structure on fatigue life of friction stir spot welding in magnesium AZ31 alloy. In: Mathaudhu SN, Sillekens WH, Neelameggham NR, Hort N, editors. Magnesium technology. Cham: Springer; 2012.

[7] Jordon JB, Horstemeyer MF, Daniewicz SR, Badarinarayan H, Grantham J. Fatigue characterization and modeling of friction stir spot welds in magnesium AZ31 alloy. J Eng Mater Technol 2010;132:041008. Available from: https://doi.org/10.1115/1.4002330.

[8] Tran V-X, Pan J, Pan T. Fatigue behavior of aluminum 5754-O and 6111-T4 spot friction welds in lap-shear specimens. Int J Fatigue 2008;30:2175—90. Available from: https://doi.org/10.1016/j.ijfatigue.2008.05.025.

[9] Moraes JFCFC, Rodriguez RII, Jordon JBB, Su X. Effect of overlap orientation on fatigue behavior in friction stir linear welds of magnesium alloy sheets. Int J Fatigue 2017;100:1—11. Available from: https://doi.org/10.1016/j.ijfatigue.2017.02.018.

[10] Fersini D, Pirondi A. Fatigue behaviour of Al2024-T3 friction stir welded lap joints. Eng Fract Mech 2007;74:468—80. Available from: https://doi.org/10.1016/j.engfracmech.2006.07.010.

[11] Bilici MK, Yükler AI. Influence of tool geometry and process parameters on macrostructure and static strength in friction stir spot welded polyethylene sheets. Mater Des 2012;33:145—52. Available from: https://doi.org/10.1016/j.matdes.2011.06.059.

[12] de Giorgi M, Scialpi A, Panella FW, de Filippis LAC. Effect of shoulder geometry on residual stress and fatigue properties of AA6082 FSW joints. J Mech Sci Technol 2009;23:26—35. Available from: https://doi.org/10.1007/s12206-008-1006-4.

[13] Rao HM, Jordon JB, Barkey ME, Guo YB, Su X, Badarinarayan H. Influence of structural integrity on fatigue behavior of friction stir spot welded AZ31 Mg alloy. Mater Sci Eng A 2013;564:369—80. Available from: https://doi.org/10.1016/j.msea.2012.11.076.

[14] De Jesus JS, Loureiro A, Costa JM, Ferreira JM. Effect of tool geometry on friction stir processing and fatigue strength of MIG T welds on Al alloys. J Mater Process Technol 2014;214:2450—60. Available from: https://doi.org/10.1016/j.jmatprotec.2014.05.012.

[15] Tran V-X, Pan J, Pan T-Y. Fatigue behavior of dissimilar 5754/7075 and 7075/5754 spot friction welds in lap-shear specimens. SAE Int J Mater Manuf 2010;3:608—14. Available from: https://doi.org/10.4271/2010-01-0961.

[16] Uematsu Y, Tokaji K, Shibata H, Tozaki Y, Ohmune T. Fatigue behaviour of friction stir welds without neither welding flash nor flaw in several aluminium alloys. Int J Fatigue 2009;31:1443—53. Available from: https://doi.org/10.1016/j.ijfatigue.2009.06.015.

[17] Badarinarayan H, Yang Q, Zhu S. Effect of tool geometry on static strength of friction stir spot-welded aluminum alloy. Int J Mach Tools Manuf 2009;49:142—8. Available from: https://doi.org/10.1016/j.ijmachtools.2008.09.004.

[18] Yin YH, Sun N, North TH, Hu SS. Influence of tool design on mechanical properties of AZ31 friction stir spot welds. Sci Technol Weld Join 2010;15:81−6. Available from: https://doi.org/10.1179/136217109X12489665059384.

[19] Longo M, D'Urso G, Giardini C, Ceretti E. Process parameters effect on mechanical properties and fatigue behavior of friction stir weld AA6060 joints. J Eng Mater Technol 2012;134:021006. Available from: https://doi.org/10.1115/1.4005916.

[20] Baragetti S, D'Urso G. Aluminum 6060-T6 friction stir welded butt joints: fatigue resistance with different tools and feed rates. J Mech Sci Technol 2014;28:867−77. Available from: https://doi.org/10.1007/s12206-013-1152-1.

[21] Ouyang JH, Kovacevic R. Material flow and microstructure in the friction stir butt welds of the same and dissimilar aluminum alloys. J Mater Eng Perform 2002;11:51−63.

[22] Cao X, Jahazi M. Effect of tool rotational speed and probe length on lap joint quality of a friction stir welded magnesium alloy. Mater Des 2011;32:1−11. Available from: https://doi.org/10.1016/j.matdes.2010.06.048.

[23] Dubourg L, Merati A, Jahazi M. Process optimisation and mechanical properties of friction stir lap welds of 7075-T6 stringers on 2024-T3 skin. Mater Des 2010;31:3324−30. Available from: https://doi.org/10.1016/j.matdes.2010.02.002.

[24] Han B, Huang Y, Lv S, Wan L, Feng J, Fu G. AA7075 bit for repairing AA2219 keyhole by filling friction stir welding. Mater Des 2013;51:25−33. Available from: https://doi.org/10.1016/j.matdes.2013.03.089.

[25] Huang YX, Han B, Lv SX, Feng JC, Liu HJ, Leng JS, et al. Interface behaviours and mechanical properties of filling friction stir weld joining AA 2219. Sci Technol Weld Join 2012;17:225−30. Available from: https://doi.org/10.1179/1362171811Y.0000000100.

[26] Li W, Li J, Zhang Z, Gao D, Wang W, Dong C. Improving mechanical properties of pinless friction stir spot welded joints by eliminating hook defect. Mater Des 2014;62:247−54. Available from: https://doi.org/10.1016/j.matdes.2014.05.028.

[27] Xu RZ, Ni DR, Yang Q, Liu CZ, Ma ZY. Pinless friction stir spot welding of Mg-3Al-1Zn alloy with Zn interlayer. J Mater Sci Technol 2016;32:76−88. Available from: https://doi.org/10.1016/j.jmst.2015.08.012.

[28] Chiou YC, Liu CTe, Lee RT. A pinless embedded tool used in FSSW and FSW of aluminum alloy. J Mater Process Technol 2013;213:1818−24. Available from: https://doi.org/10.1016/j.jmatprotec.2013.04.018.

[29] Tozaki Y, Uematsu Y, Tokaji K. A newly developed tool without probe for friction stir spot welding and its performance. J Mater Process Technol 2010;210:844−51. Available from: https://doi.org/10.1016/j.jmatprotec.2010.01.015.

[30] Uematsu Y, Tokaji K, Tozaki Y, Kurita T, Murata S. Effect of re-filling probe hole on tensile failure and fatigue behaviour of friction stir spot welded joints in Al-Mg-Si alloy. Int J Fatigue 2008;30:1956−66. Available from: https://doi.org/10.1016/j.ijfatigue.2008.01.006.

[31] Rao HM, Jordon JB, Boorgu SK, Kang H, Yuan W, Su X. Influence of the key-hole on fatigue life in friction stir linear welded aluminum to magnesium. Int J Fatigue 2017;105:16−26. Available from: https://doi.org/10.1016/j.ijfatigue.2017.08.012.

[32] Rao HM, Ghaffari B, Yuan W, Jordon JB, Badarinarayan H. Effect of process parameters on microstructure and mechanical behaviors of friction stir linear welded aluminum to magnesium. Mater Sci Eng A 2016;651:27−36. Available from: https://doi.org/10.1016/j.msea.2015.10.082.

[33] James MN, Hattingh DG, Bradley GR. Weld tool travel speed effects on fatigue life of friction stir welds in 5083 aluminium. Int J Fatigue 2003;25:1389−98. Available from: https://doi.org/10.1016/S0142-1123(03)00061-6.

[34] Rodriguez RI, Jordon JB, Rao HM, Badarinarayan H, Yuan W, El Kadiri H, et al. Microstructure, texture, and mechanical properties of friction stir spot welded rare-earth containing ZEK100 magnesium alloy sheets. Mater Sci Eng A 2014;618:637–44. Available from: https://doi.org/10.1016/j.msea.2014.09.010.

[35] Rao HM, Yuan W, Badarinarayan H. Effect of process parameters on mechanical properties of friction stir spot welded magnesium to aluminum alloys. Mater Des 2015;66:235–45. Available from: https://doi.org/10.1016/J.MATDES.2014.10.065.

[36] Naik BS, Chen DL, Cao X, Wanjara P. Microstructure and fatigue properties of a friction stir lap welded magnesium alloy. Metall Mater Trans A Phys Metall Mater Sci 2013;44:3732–46. Available from: https://doi.org/10.1007/s11661-013-1728-5.

[37] Rao HM, Rodriguez RI, Jordon JB, Barkey ME, Guo YB, Badarinarayan H, et al. Friction stir spot welding of rare-earth containing ZEK100 magnesium alloy sheets. Mater Des 2014;56:750–4. Available from: https://doi.org/10.1016/j.matdes.2013.12.034.

[38] Yuan W, Carlson B, Verma R, Szymanski R. Study of top sheet thinning during friction stir lap welding of AZ31 magnesium alloy. Sci Technol Weld Join 2012;17:375–80. Available from: https://doi.org/10.1179/1362171812Y.0000000018.

[39] Lathabai S, Painter MJ, Cantin GMD, Tyagi VK. Friction spot joining of an extruded Al–Mg–Si alloy. Scr Mater 2006;55:899–902. Available from: https://doi.org/10.1016/j.scriptamat.2006.07.046.

[40] Shin H-S, Jung Y-C, Lee J-K. Influence of tool speeds on dissimilar friction stir spot welding characteristics of bulk metallic glass/Mg alloy. Met Mater Int 2012;18:685–9. Available from: https://doi.org/10.1007/s12540-012-4018-7.

[41] Dehghani M, Amadeh A, Akbari Mousavi SAA. Investigations on the effects of friction stir welding parameters on intermetallic and defect formation in joining aluminum alloy to mild steel. Mater Des 2013;49:433–41. Available from: https://doi.org/10.1016/j.matdes.2013.01.013.

[42] Kimapong K, Watanabe T. Effect of welding process parameters on mechanical property of FSW lap joint between aluminum alloy and steel. Mater Trans 2005;46:2211–17. Available from: https://doi.org/10.2320/matertrans.46.2211.

[43] Bahrami M, Helmi N, Dehghani K, Givi MKB. Exploring the effects of SiC reinforcement incorporation on mechanical properties of friction stir welded 7075 aluminum alloy: fatigue life, impact energy, tensile strength. Mater Sci Eng A 2014;595:173–8. Available from: https://doi.org/10.1016/j.msea.2013.11.068.

[44] Seighalani KR, Givi MKB, Nasiri AM, Bahemmat P. Investigations on the effects of the tool material, geometry, and tilt angle on friction stir welding of pure titanium. J Mater Eng Perform 2010;19:955–62. Available from: https://doi.org/10.1007/s11665-009-9582-8.

[45] Elyasi M, Aghajani Derazkola H, Hosseinzadeh M. Investigations of tool tilt angle on properties friction stir welding of A441 AISI to AA1100 aluminium. Proc Inst Mech Eng Part B J Eng Manuf 2016;230:1234–41. Available from: https://doi.org/10.1177/0954405416645986.

[46] Yadava MK, Mishra RS, Chen YL, Carlson B, Grant GJ. Study of friction stir joining of thin aluminium sheets in lap joint configuration. Sci Technol Weld Join 2010;15:70–5. Available from: https://doi.org/10.1179/136217109X12537145658733.

[47] Cederqvist L, Reynolds AP. Factors affecting the properties of friction stir welded aluminum lap joints. Weld J 2001;80:281–7.

[48] Yang Q, Li X, Chen K, Shi YJ. Effect of tool geometry and process condition on static strength of a magnesium friction stir lap linear weld. Mater Sci Eng A 2011;528:2463–78. Available from: https://doi.org/10.1016/j.msea.2010.12.030.

[49] Cao XJ, Jahazi M. Friction stir welding of dissimilar AA 2024-T3 to AZ31B-H24 alloys. Mater Sci Forum 2010;638–642:3661–6. Available from: https://doi.org/10.4028/www.scientific.net/MSF.638-642.3661.

[50] Firouzdor V, Kou S. Al-to-Mg friction stir welding: Effect of material position, travel speed, and rotation speed. Metall Mater Trans A 2010;41:2914–35. Available from: https://doi.org/10.1007/s11661-010-0340-1.

[51] Yan J, Xu Z, Li Z, Li L, Yang S. Microstructure characteristics and performance of dissimilar welds between magnesium alloy and aluminum formed by friction stirring. Scr Mater 2005;53:585–9. Available from: https://doi.org/10.1016/j.scriptamat.2005.04.022.

[52] Fu B, Qin G, Li F, Meng X, Zhang J, Wu C. Friction stir welding process of dissimilar metals of 6061-T6 aluminum alloy to AZ31B magnesium alloy. J Mater Process Technol 2015;218:38–47. Available from: https://doi.org/10.1016/j.jmatprotec.2014.11.039.

[53] Yamamoto N, Liao J, Watanabe S, Nakata K. Effect of intermetallic compound layer on tensile strength of dissimilar friction-stir weld of a high strength Mg alloy and Al alloy. Mater Trans 2009;50:2833–8. Available from: https://doi.org/10.2320/matertrans.M2009289.

[54] Sahu PK, Pal S, Pal SK, Jain R. Influence of plate position, tool offset and tool rotational speed on mechanical properties and microstructures of dissimilar Al/Cu friction stir welding joints. J Mater Process Technol 2016;235:55–67. Available from: https://doi.org/10.1016/j.jmatprotec.2016.04.014.

[55] Tan CW, Jiang ZG, Li LQ, Chen YB, Chen XY. Microstructural evolution and mechanical properties of dissimilar Al-Cu joints produced by friction stir welding. Mater Des 2013;51:466–73. Available from: https://doi.org/10.1016/j.matdes.2013.04.056.

[56] Xue P, Ni DR, Wang D, Xiao BL, Ma ZY. Effect of friction stir welding parameters on the microstructure and mechanical properties of the dissimilar Al-Cu joints. Mater Sci Eng A 2011;528:4683–9. Available from: https://doi.org/10.1016/j.msea.2011.02.067.

[57] Rao HM, Jordon JB, Ghaffari B, Su X, Khosrovaneh AK, Barkey ME, et al. Fatigue and fracture of friction stir linear welded dissimilar aluminum-to-magnesium alloys. Int J Fatigue 2016;82:737–47.

[58] Richter-Trummer V, Suzano E, Beltrão M, Roos A, dos Santos JF, de Castro PMST. Influence of the FSW clamping force on the final distortion and residual stress field. Mater Sci Eng A 2012;538:81–8. Available from: https://doi.org/10.1016/j.msea.2012.01.016.

[59] Reynolds AP, Tang W, Gnaupel-Herold T, Prask H. Structure, properties, and residual stress of 304L stainless steel friction stir welds. Scr Mater 2003;48:1289–94. Available from: https://doi.org/10.1016/S1359-6462(03)00024-1.

[60] Lombard H, Hattingh DG, Steuwer A, James MN. Effect of process parameters on the residual stresses in AA5083-H321 friction stir welds. Mater Sci Eng A 2009;501:119–24. Available from: https://doi.org/10.1016/j.msea.2008.09.078.

[61] Hatamleh O, Rivero IV, Swain SE. An investigation of the residual stress characterization and relaxation in peened friction stir welded aluminum-lithium alloy joints. Mater Des 2009;30:3367–73. Available from: https://doi.org/10.1016/j.matdes.2009.03.038.

[62] Peel M, Steuwer A, Preuss M, Withers PJ. Microstructure, mechanical properties and residual stresses as a function of welding speed in aluminium AA5083 friction stir welds. Acta Mater 2003;51:4791–801. Available from: https://doi.org/10.1016/S1359-6454(03)00319-7.

[63] Jamshidi Aval H. Microstructure and residual stress distributions in friction stir welding of dissimilar aluminium alloys. Mater Des 2015;87:405–13. Available from: https://doi.org/10.1016/j.matdes.2015.08.050.

[64] Brewer LN, Bennett MS, Baker BW, Payzant EA, Sochalski-Kolbus LM. Characterization of residual stress as a function of friction stir welding parameters in oxide dispersion strengthened (ODS) steel MA956. Mater Sci Eng A 2015;647:313–21. Available from: https://doi.org/10.1016/j.msea.2015.09.020.

[65] James MN, Hughes DJ, Hattingh DG, Bradley GR, Mills G, Webster PJ. Synchrotron diffraction measurement of residual stresses in friction stir welded 5383-H321 aluminium butt

joints and their modification by fatigue cycling. Fatigue Fract Eng Mater Struct 2004;27:187−202. Available from: https://doi.org/10.1111/j.1460-2695.2004.00736.x.

[66] Solanki KN, Jordon JB, Whittington W, Rao H, Hubbard CR. Structure−property relationships and residual stress quantification of a friction stir spot welded magnesium alloy. Scr Mater 2012;66:797−800. Available from: https://doi.org/10.1016/j.scriptamat.2012.02.011.

[67] Sun T, Roy MJ, Strong D, Withers PJ, Prangnell PB. Comparison of residual stress distributions in conventional and stationary shoulder high-strength aluminum alloy friction stir welds. J Mater Process Technol 2017;242:92−100. Available from: https://doi.org/10.1016/j.jmatprotec.2016.11.015.

[68] Zapata J, Toro M, López D. Residual stresses in friction stir dissimilar welding of aluminum alloys. J Mater Process Technol 2016;229:121−7. Available from: https://doi.org/10.1016/j.jmatprotec.2015.08.026.

[69] Dorman M, Toparli MB, Smyth N, Cini A, Fitzpatrick ME, Irving PE. Effect of laser shock peening on residual stress and fatigue life of clad 2024 aluminium sheet containing scribe defects. Mater Sci Eng A 2012;548:142−51. Available from: https://doi.org/10.1016/j.msea.2012.04.002.

[70] Hatamleh O. A comprehensive investigation on the effects of laser and shot peening on fatigue crack growth in friction stir welded AA 2195 joints. Int J Fatigue 2009;31:974−88. Available from: https://doi.org/10.1016/j.ijfatigue.2008.03.029.

[71] Sharma V, Sharma C, Upadhyay V, Singh S. Enhancing mechanical properties of friction stir welded joints of Al-Si-Mg alloy through post weld heat treatments. Mater Today Proc 2017;4:628−36. Available from: https://doi.org/10.1016/j.matpr.2017.01.066.

[72] Elangovan K, Balasubramanian V. Influences of post-weld heat treatment on tensile properties of friction stir-welded AA6061 aluminum alloy joints. Mater Charact 2008;59:1168−77. Available from: https://doi.org/10.1016/j.matchar.2007.09.006.

[73] Bayazid SM, Farhangi H, Asgharzadeh H, Radan L, Ghahramani A, Mirhaji A. Effect of cyclic solution treatment on microstructure and mechanical properties of friction stir welded 7075 Al alloy. Mater Sci Eng A 2016;649:293−300. Available from: https://doi.org/10.1016/j.msea.2015.10.010.

[74] Safarbali B, Shamanian M, Eslami A. Effect of post-weld heat treatment on joint properties of dissimilar friction stir welded 2024-T4 and 7075-T6 aluminum alloys. Trans Nonferrous Met Soc China 2018;28:1287−97. Available from: https://doi.org/10.1016/S1003-6326(18)64766-1.

[75] Aydin H, Bayram A, Uğuz A, Akay KS. Tensile properties of friction stir welded joints of 2024 aluminum alloys in different heat-treated-state. Mater Des 2009;30:2211−21. Available from: https://doi.org/10.1016/j.matdes.2008.08.034.

[76] Costa MI, Leitão C, Rodrigues DM. Influence of post-welding heat-treatment on the monotonic and fatigue strength of 6082-T6 friction stir lap welds. J Mater Process Technol 2017;250:289−96. Available from: https://doi.org/10.1016/j.jmatprotec.2017.07.030.

[77] White BC, Rodriguez RI, Cisko A, Jordon JB, Allison PG, Rushing T, et al. Effect of heat exposure on the fatigue properties of AA7050 friction stir welds. J Mater Eng Perform 2018;. Available from: https://doi.org/10.1007/s11665-018-3379-6.

CHAPTER 4

Fatigue Crack Growth in Friction Stir Welds

4.1 INTRODUCTION TO FATIGUE CRACK GROWTH CONCEPTS

The use of fatigue crack growth (FCG) data and concepts is common in life estimation for structures in which damage-tolerant design is applicable and in particular for friction stir welding (FSW). Welded structures that may be candidates for damage-tolerant design include pipelines, civil infrastructure such as bridges and buildings, airframes, and automobile chassis. One might say that, in general, damage-tolerant design can be applied to structures joined with FSW, which are made of ductile materials that experience global loading conditions resulting in relatively high factors of safety with respect to stress. Specifically, damage-tolerant design methods for FSW are generally most applicable under the following conditions: (1) the FSW are routinely inspected for cracks, (2) inspection methods are capable of detecting cracks of appropriately small sizes and can detect said cracks with certainty, (3) crack extension can be monitored, (4) crack extension occurs in a controlled, systematic, fashion such that a crack may grow at a predictable rate, (5) there remains sufficient useful life during crack growth from detected size to critical crack length, and (6) failure by critical crack size results in non-catastrophic events. Inherent to the six conditions listed above is the understanding that the FSW joint in question exhibits ductile FCG characteristics, and the FCG response of the FSW joint is well understood.

The FCG materials or structures having varying microstructures add another level of complexity to the application of damage-tolerant design. Microstructural features that may affect FCG response include grain size, grain texture, microstructural phases present (e.g., ferrite vs martensite in steels), and strengthening phases present (e.g., precipitates). As in fusion-based welding, FSW of initially homogeneous microstructures will result in a graded microstructure which alters the FCG response. Specifically, the FSW process may have produced sufficient heat as to induce a phase change within the nugget of the FSW. The FSW will also likely exhibit relatively small, equiaxed, grain

Fatigue in Friction Stir Welding. DOI: https://doi.org/10.1016/B978-0-12-816131-9.00004-0
© 2019 Elsevier Inc. All rights reserved.

structure when compared to the base material (BM). The thermome-chanically affected zone (TMAZ) of the weld will likely exhibit relatively elongated grains and fairly strong texture resulting from the large-scale shear induced during the process. Finally, the nugget, the TMAZ, and the heat affected zone (HAZ) of the FSW may have experienced sufficient temperatures during welding which resulted in coarsening or complete dissolution of strengthening precipitates. In short, the FSW process may alter all microstructural aspects that affect FCG response of a particular material. However, we note that caution should be exercised when applying damage-tolerant design to components fabricated using FSW. Although the crack path of least resistance in a homogeneous microstructure is strongly dominated by the orientation of external loads, the crack path in the graded microstructures produced during FSW is also strongly dominated by the microstructural features.

The application of FCG data to damage-tolerant design of FSW joints may also be complicated by the existence of multiaxial loading, difficulty in quantifying time-varying loading conditions, nontrivial boundary conditions resulting in crack tip constraints, environmental conditions, and varying cross-sectional geometry (among others) within the welded structure. As a minimum criterion, if the plastic zone produced at the crack tip within the FSW is small relative to the crack length and structural geometry, the concept of similitude may be assumed to be applicable and FCG data collected per appropriate standards on laboratory specimens can be applied to the analysis of the structure. The concept of similitude, which is a critical foundation of linear elastic fracture mechanics (LEFM), is the science supporting the applicability of damage-tolerant design. Again, realistic loading conditions, boundary conditions, and environmental conditions must be accounted for if one attempts to use FCG data for damage-tolerant design in FSW. This chapter will introduce these concepts in more detail and provide applications to FSW metals. In addition, this chapter will provide a brief discussion on the effect of residual stresses on FCG. Lastly, characterization of fatigue mechanisms including FCG in FSSW is presented.

4.1.1 Loading Modes
The load application in nearly every realistic scenario, including loading in FSW, may be decomposed into contributions from one, or

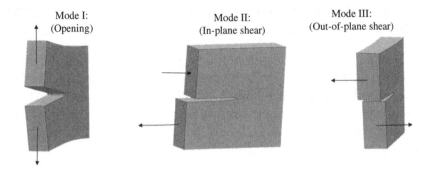

Figure 4.1 Representation of potential loading modes present in structures.

multiple, of three primary loading modes. The loading modes are designated as Mode I, Mode II, and Mode III. Fig. 4.1 depicts the loading modes applied to a generic body having a crack.

The most common FCG test conducted on FSW joints, regardless of material, is performed by use of Mode I loading. Mode I loading is characterized by the application of loads perpendicular to the crack face and therefore the crack experiences pure opening displacements. Mode II loading produces crack surfaces that slide past one-another in the direction of loading. Mode III loading results in a tearing action that moves the crack surfaces in opposite directions, yet coplanar with the crack propagation plane. Although Mode I loading results in direct opening of the crack tip, Mode II and Mode III result in pure tearing (or shear) with (theoretically) no resultant crack tip opening.

4.1.2 Stress Intensity Factor

A crucial concept to discuss with respect to the FCG of FSWs is the concept of a stress intensity. In general, a crack tip within a continuum of media behaves as a singularity with respect to stress. That is, the stress magnitude, σ, goes as a function of $\sigma \propto 1/\sqrt{r}$. This relationship dictates that, in theory, the stress magnitude approaches infinity as the distance in front of the crack tip, r, reduces to zero. One convenient way to define the crack tip singularity is by the use of the stress intensity factor. The stress intensity factor is defined as

$$K = \beta\sigma\sqrt{2\pi r},\qquad(4.1)$$

where K is the stress intensity factor; β is a factor that accounts for crack length (a), geometry, loading conditions, and boundary conditions; σ is a stand-in for a generic nominal stress; and r is the distance in front of the crack tip.

If geometry and loading conditions produce Mode I loading, the resultant stress intensity factor would be termed K_I, or the Mode I stress intensity factor. This concept extends to Mode II and Mode III loading as well. Finally, if the loading is cyclic, the stress intensity factor range, $\Delta K = K_{max} - K_{min}$, would be defined as a function of the applied stress range, $\Delta \sigma = \sigma_{max} - \sigma_{min}$, as

$$\Delta K = \beta \Delta \sigma \sqrt{2\pi r}. \tag{4.2}$$

In order to determine the value of the stress intensity factor, or the stress intensity factor range, one must have a closed-form solution for the parameter β. There are handbooks of tabulated solutions for the stress intensity factor, and therefore β, for most laboratory specimens of interest. The most common laboratory specimen to perform FCG experiments on FSW joints is the compact tension (CT) specimen, although other specimens like middle-tension have been used [1]. The closed-form solution for the CT specimen provided in ASTM E 647 is as follows:

$$\Delta K = \frac{\Delta P}{B\sqrt{W}} \frac{(2+\alpha)}{(1-\alpha)^{3/2}} \left(0.886 + 4.64\alpha - 13.32\alpha^2 + 14.72\alpha^3 - 5.6\alpha^4\right).$$

$$\tag{4.3}$$

where ΔP is the applied load range defined as $\Delta P = P_{max} - P_{min}$, B is the CT specimen thickness, W is the characteristic width of the CT specimen (defined as the distance from the load line to the back edge), $\alpha = a/W$, and a is the crack length measured from the load line.

The application of Eq. 4.3 is limited to linear elastic, isotropic, and homogeneous materials; specimen geometries that meet the ASTM 647 criteria; crack lengths larger than $a/W \geq 0.2$; and that LEFM assumptions are not invalidated (e.g., the specimen remains primarily elastic in its stress response). One will note when comparing Eqs. 4.2 and 4.3 that the β-term is nontrivial, even for the (seemingly) trivial CT specimen geometry. It can then be understood that closed-form solutions for the stress intensity factor for "realistic" geometries, loading conditions, and boundary conditions encountered in the field can sometimes not exist.

An interesting corollary to Eq. 4.1 is discussed next. If one knows the value of the stress intensity factor, one can then determine the state of stress at a crack tip. The relationships for the full state of stress at a crack tip experiencing Mode I loading are as follows:

$$\sigma_{xx} = \frac{K_I}{\sqrt{2\pi r}} \cos\left(\frac{\theta}{2}\right) \left[1 - \sin\left(\frac{\theta}{2}\right) \sin\left(\frac{3\theta}{2}\right)\right], \qquad (4.4)$$

$$\sigma_{yy} = \frac{K_I}{\sqrt{2\pi r}} \cos\left(\frac{\theta}{2}\right) \left[1 + \sin\left(\frac{\theta}{2}\right) \sin\left(\frac{3\theta}{2}\right)\right], \qquad (4.5)$$

$$\sigma_{zz} = \begin{cases} 0, \text{for plane stress} \\ v(\sigma_{xx} + \sigma_{yy}), \text{for plane strain} \end{cases}, \qquad (4.6)$$

$$\sigma_{xy} = \frac{K_I}{\sqrt{2\pi r}} \cos\left(\frac{\theta}{2}\right) \sin\left(\frac{\theta}{2}\right) \cos\left(\frac{3\theta}{2}\right), \qquad (4.7)$$

$$\sigma_{xz} = \sigma_{yz} = 0 \qquad (4.8)$$

The concept of similitude was mentioned in the introductory section of this chapter. When applied to LEFM, the concept of similitude dictates that the stress intensity factor is size and geometry independent, and therefore laboratory results are directly applicable to structures. Similitude is a necessary, although not sufficient, criterion to be met to apply LEFM appropriately. In order to ensure similitude, the FSW joint must be linear elastic, the majority of the FSW and surrounding BM must remain linear elastic during loading, and the crack must be small relative to the overall geometry. The second condition is met when the plastic zone at the crack tip is small compared to both the crack length and the overall geometry. This criterion is explicitly met when the plastic zone is completely contained within what is called the singularity dominated zone. Unfortunately, there are no hard and fast rules defining the size of the singularity dominated zone, especially in FSW. Furthermore, the size of the plastic zone ahead of a crack tip is estimated as a function of the yield stress. Given that the FSW process produces a graded microstructure, the plastic zone size will likely vary as the crack progresses through different regions of the weld. As an example, if a crack were to initiate in the BM or nugget, the crack may transit to the region between the nugget and TMAZ. In doing so the plastic zone size will also change as a function of the modified material microstructures encountered. While the transition in crack path will result in varying FCG rates (FCGRs), it may also affect the applicability of similitude and therefore negate one's ability to apply LEFM. Caution is suggested here: even when similitude is satisfied, one may never find a suitable closed-form solution for the stress intensity factor of the FSW joint and or structure of interest.

Laboratory tests on FSW joints are typically conducted by use of a single mode of loading. As mentioned above, the most common laboratory test to produce FCG and fracture data of FSW is performed on CT specimens in Mode I loading. However, it is very common for structures to experience mixed-mode loading, especially in lap-shear coupons as described in Chapter 2, Fatigue Behavior in Friction Stir Welds. Mixed-mode loading is one in which any combination of Mode I/Mode II/Mode III loads are applied simultaneously. In this case one must realize that although stress tensors resulting from different loading modes may be summed, stress intensity factors from different loading modes may not be summed. Specifically, the following relationship for stress is valid:

$$\sigma_{ij}^{Total} = \sigma_{ij}^{Mode\ I} + \sigma_{ij}^{Mode\ II} + \sigma_{ij}^{Mode\ III}. \tag{4.9}$$

One will note the inequality sign on the relationship provided in Eq. 4.10, as summation of stress intensity factors produced from different modes is not valid.

$$K_{Total} \neq K_I + K_{II} + K_{III} \tag{4.10}$$

4.1.3 FCG Curves

In order to perform damage-tolerant design for structures fabricated using FSW, one must have applicable FCG data available for analysis such as material-specific, loading-specific, and environment-specific data. FCG data are created by use of a considerable amount of testing. For ease of discussion, and without loss of generality, we will assume that a test regimen has been set up to create FCG data on a CT specimen for a butt-friction stir welding (FSBW). The assumed test regimen will progress by application of load (as opposed to displacement), and the load waveform will be maximum-to-zero ($P_{min} = 0$). The load applied to the specimen as a function of time, $P(t)$, and the crack length as a function of cycles, $a(N)$, is collected as the test progresses. There are many ways to determine the crack length and the interested reader is directed to the appropriate ASTM standards for more

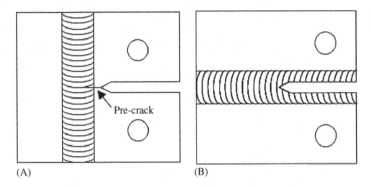

Figure 4.2 Example of a compact tension oriented transverse (A) and longitudinal (B) to the weld zone [2]. Reprinted with permission from John Wiley and Sons.

information. It is important to note that another variable to consider in the FCG testing of FSW is the orientation of the weld relative to the applied load. Fig. 4.2 shows an example of CT with the FSW orientated parallel and perpendicular to the notch. The CT specimens shown in Fig. 4.2 illustrate that tests can be used to measure the FCG within the weld zone or measure the FCG as the crack propagates across the FSW.

Given that P_{max} and P_{min} are constant for all time, we have a single ΔP for a particular data set. Having collected $a(N)$ and knowing ΔP, one can calculate the stress intensity factor range for all crack lengths occurring in the test, $\Delta K(a)$, by use of Eq. 4.3. Additionally, one may determine the FCGR, da/dN, by taking the derivative of the a versus N data for all data collected. Note by definition the FCGR, da/dN, is the crack extension per cycle, for all cycles occurring in the test. Finally, one can plot the calculated da/dN values versus the associated calculated ΔK values on log–log axis to create the FCG curve. The FCG plot is often referred to as a da/dN-ΔK plot

When plotting da/dN versus ΔK, three distinct regions are manifested within the data set. As shown in Fig. 4.3, the different FCG regions are termed Region I, Region II, and Region III. Region I is also known as the threshold or small crack growth region. The threshold stress intensity range, ΔK_{th}, is the stress intensity range below which a crack will not grow. If a small crack exists within the FSW, and the stress intensity range for that crack is above the threshold value, the crack will likely grow erratically within Region I. This is due to the crack length being smaller than, or on the order of,

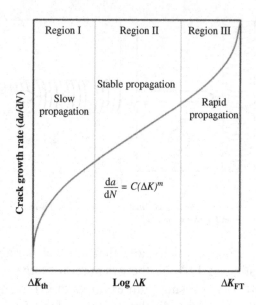

Figure 4.3 Regions of FCG in a da/dN versus ΔK plot [3]. Reprinted with permission from Springer Nature.

microstructural features as well as the plastic zone created by the crack. As such, crack growth within Region I is highly sensitive to microstructure and the post-yielding deformation response of the material. As a result, crack propagation and arrest often manifest in Region I FCG data and exhibit no clear trend on the da/dN versus ΔK curve. This is of particular interest when applying LEFM to graded microstructures such as those found in FSW. The varying microstructures will greatly affect the crack initiation and microstructurally small crack growth response of the structure. A crack traversing from one microstructure in the FSW to another different microstructure in a FSW joint could result in a drastic increase in the FCGR or crack arrest. The da/dN-ΔK data in Region II, sometimes referred to as the Paris region, often exhibits a linear trend when plotted on the log–log axis. FCG in Region II results from cracks that are large relative to the grain size of the material and therefore is typically insensitive to that aspect of material microstructure. Furthermore, the FCGR in this region is often predictable. Therefore, Region II FCG is ideally suited for damage-tolerant design of structure joined by FSW. Region III, often referred to as the fast crack growth region, manifests in a sharp upwards turn in the da/dN-ΔK data. This region ends at specimen fracture/separation. The manifestation of fast fracture is sensitive to microstructure in that a ductile material will exhibit micro-void growth and coalescence failure mechanisms, whereas a more brittle material will exhibit cleavage failure mechanisms in this regime.

Given that FCG data within Region II exhibits a linear trend when plotted on a log–log scale, the data within this region are tabulated and fit to a power law relationship:

$$\frac{da}{dN} = c\Delta K^m \tag{4.11}$$

where c and m are material-specific and environment-specific constants. Eq. 4.11 is known as the Paris law or Paris Equation, and tabulated values of c and m are available for many materials of interest, particularly in ambient environments. Care must be taken when selecting Paris constants for FSW materials, as the FCG response is specific to the weld region within which the crack is growing. Given that FCGRs within the Paris region are stable and predictable for most materials in ambient environments, this region of FCG is perfectly suited for damage-tolerant design. However, in many cases, Paris constants for an FSW of particular material may not be available. However, studies have shown generally acceptable life predictions using Paris constants for BMs when values for FSW are not available [4–6]. Further discussion of damage-tolerant design in FSW is presented in Chapter 5, Fatigue Modeling of Friction Stir Welding.

4.1.4 Mean Stress Effects and Crack Closure

Often times the load experienced by a crack is not fully reversed. In which case, the crack tip will experience a nonzero mean load, and therefore a nonzero mean stress. The load ratio may be defined as

$$R = \frac{P_{min}}{P_{max}} = \frac{K_{min}}{K_{max}} = \frac{\sigma_{min}}{\sigma_{max}} = \frac{\varepsilon_{min}}{\varepsilon_{max}} = \frac{\delta_{min}}{\delta_{max}}, \tag{4.12}$$

depending upon the type of "load" that is being applied to the structure.

The load ratio is often used in conjunction with ΔK and/or ΔP to fully describe the external loading applied to a body experiencing fatigue. The FCG data found in the literature will often be created by use of the following load ratios: $R = 0.5$ or $R = 0.05$ ($R \sim 0$). A positive load ratio will produce a positive mean stress, and a negative load ratio may produce a negative, positive, or zero mean stress. For all other variables held equal, a more positive mean stress will increase the FCGR exhibited by a material. This effect is shown in Fig. 4.4, where the FCG for an aluminum alloy 6061 subjected to various load ratios exhibited higher crack growth rates for more positive ratios and lower crack growth rates for less positive ratios.

The effect of changing load ratio can be accounted for by considering several relationships. A simple, well established FCG relationship incorporating variable load ratio was proposed by Walker and is provided below

$$\frac{da}{dN} = c[(1-R)^n K_{max}]^m \tag{4.13}$$

where the constants c and m are those provided in Eq. 4.11 and therefore may be found tabulated in the literature, and n is fit to experimental data. Note that all constants in Eq. 4.13 are environment specific.

Crack closure is a phenomenon that may occur during FCG of FSW that causes a retardation of the crack growth rate, for all other variables held equal. Crack closure typically manifests in a compressive residual stress applied to the crack tip that must be overcome prior to the physical opening of the crack tip. One may notice the crack closure mechanism occurring when one reduces the opening load on a cracked specimen. That is, if the crack closes before fully unloading the

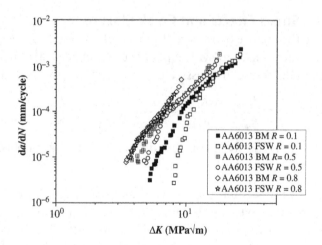

Figure 4.4 Fatigue crack growth data for AA6013 BM and FSW as a function of load ratio. Data replotted from Biallas G. Effect of welding residual stresses on fatigue crack growth thresholds. Int J Fatigue 2013;50:10−7. doi:10.1016/j.ijfatigue.2012.07.002, [7].

specimen, the crack may be experiencing closure effects. Crack closure may occur as a result of multiple factors, including crack tip bridging by foreign debris, crack tip bridging by crack face interactions, and the plastic zone wake behind the crack tip. The plastic zone wake behind the crack tip is strongly affected by the size of the plastic zone in front of the crack tip and therefore the yield stress of the material. In which case, crack closure effects in graded microstructures like FSW may be nontrivial to quantify. In order to quantify the effects of crack closure in FSW experimentally, one must determine the load, and ultimately the stress intensity (K_{op}), required to open the crack tip. Note again that this value will likely change as a function of microstructure. In doing so, one may apply the FCG equation modified by Elber for crack closure provided below:

$$\Delta K_{eff} = K_{max} - K_{op} \tag{4.14}$$

$$\frac{da}{dN} = c_1 \Delta K_{eff}^m \tag{4.15}$$

Note that the constant c_1 in Eq. 4.15 is not the same value as that in Eq. 4.11 and therefore will not likely be found in tabulated data.

4.2 FRICTION STIR WELD FCG BEHAVIOR

In order to provide some examples of the FCG in FSW, representative data sets of FSW stir zones (FSW or Weld), FSW HAZ, and BM are provided in this section. The FCG data presented in this section are in addition to the FCG in the appendix and is presented here for clarity of discussion. Note that the vast majority of the data in this section were created by use of a CT specimen, in air, at a load ratio of $R = 0.1$.

4.2.1 2xxx Series Aluminums

The 2xxx-series aluminums are precipitate hardened metals. In which case the FSW process parameters, and resulting heat input, will greatly affect the precipitates present and the resulting FCGR. The data collected from the literature, and provided in this section, provide particular insight to the precipitate strengthening effect on FCG (Fig. 4.5).

One will note that while the FCG response of the BM and weld in the AA2195 and AA2024 data presented here generally trend together,

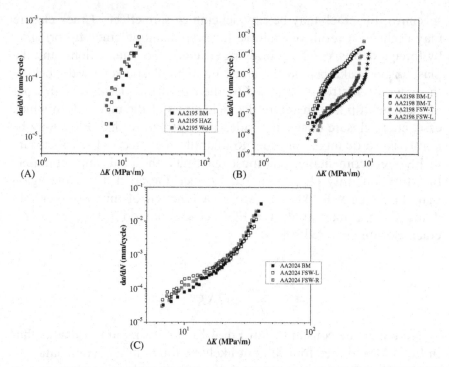

Figure 4.5 (A) FCG data for AA2195 BM, HAZ, and welds. (B) FCG data for AA2198 BM and FSW and BM as a function of orientation. (C) FCG data for AA2024 BM and FSW. (A) Data replotted from Moreira PMGP, de Jesus AMP, de Figueiredo MAV, Windisch M, Sinnema G, de Castro PMST. Fatigue and fracture behaviour of friction stir welded aluminium-lithium 2195. Theor Appl Fract Mech 2012;60:1–9. doi:10.1016/j. tafmec.2012.06.001, [8]. (B) Data replotted from Cavaliere P, De Santis A, Panella F, Squillace A. Effect of anisotropy on fatigue properties of 2198 Al-Li plates joined by friction stir welding. Eng Fail Anal 2009;16:1856–65. doi:10.1016/j.engfailanal.2008.09.024, [9]. (C) Data replotted from Okada T, Machida S, Nakamura T, Tanaka H, Kuwayama K, Asakawa M. Fatigue crack growth of friction-stir-welded aluminum alloy. J Aircr 2017;54:737–46. doi:10.2514/1.C034119, [10].

the FCG response of the BM and weld of the AA2198 data presented are vastly different. This may be a function of the three different alloys' response to optimal FSW process parameters, a result of non-optimal process parameters affecting the strengthening phase within the alloy, or something else entirely. Again, care must be taken when applying damage-tolerant design to friction stir welded structures as graded microstructures may strongly affect the FCG response. Data indicative of load ratio effects in both BMs and friction stir welds are provided below (Fig. 4.6).

As mentioned above, weld process parameters will have a primary effect on the resulting microstructures in the weld nugget, HAZ, and TMAZ. The effect of varying weld process parameters is shown in

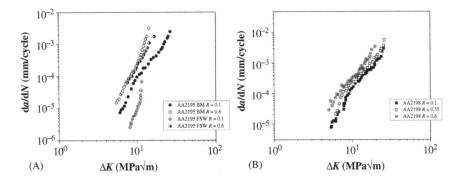

Figure 4.6 (A) FCG data for AA2195 BM and FSW as a function of load ratio. (B) FCG data for AA2198 BM as a function of load ratio. (A) Data replotted from Ma YE, Staron P, Fischer T, Irving PE. Size effects on residual stress and fatigue crack growth in friction stir welded 2195-T8 aluminium—Part I: Experiments. Int J Fatigue 2011;33:1417–25. doi:10.1016/j.ijfatigue.2011.05.006. (B) Data replotted from Ma YE, Zhao ZQ, Liu BQ, Li WY. Mechanical properties and fatigue crack growth rates in friction stir welded nugget of 2198-T8 Al-Li alloy joints. Mater Sci Eng A 2013;569:41–7. doi:10.1016/j.msea.2013.01.044, [12].

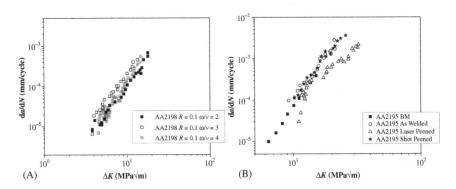

Figure 4.7 (A) FCG data for AA2198 FSW as a function of weld process parameters. (B) FCG data for AA2195 BM, as-welded weld, laser peened weld, and shot peened weld. (A) Data replotted from Ma YE, Xia ZC, Jiang RR, Li WY. Effect of welding parameters on mechanical and fatigue properties of friction stir welded 2198 T8 aluminum–lithium alloy joints. Eng Fract Mech 2013;114:1–11. doi:10.1016/j.engfracmech.2013.10.010, [13]. (B) Data replotted from Moreira PMGP, de Jesus AMP, de Figueiredo MAV, Windisch M, Sinnema G, de Castro PMST. Fatigue and fracture behaviour of friction stir welded aluminium-lithium 2195. Theor Appl Fract Mech 2012;60:1–9. doi:10.1016/j.tafmec.2012.06.001; Hatamleh O. A comprehensive investigation on the effects of laser and shot peening on fatigue crack growth in friction stir welded AA 2195 joints. Int J Fatigue 2009;31:974–88. doi:10.1016/j.ijfatigue.2008.03.029, [8,14].

Fig. 4.7. Additionally, weld postprocessing may mitigate any deleterious effects of the FSW process. Postprocessing may be in the form of heat treatment or a surface treatment such as peening. Peening introduces beneficial surface residual stresses which aid in enhancing the FCG response of the material. Data suggesting that friction stir welds may benefit from laser peening in particular are provided in Fig. 4.7B.

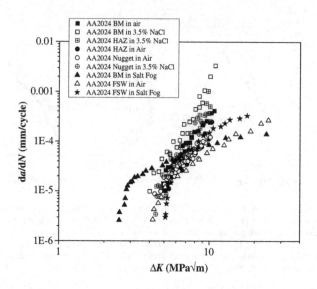

Figure 4.8 FCG data for AA2024 BM, weld, and HAZ as a function of environment. Data replotted from Wang W, Qiao K, Wu JL, Li TQ, Cai J, Wang KS. Fatigue properties of friction stir welded joint of ultrafine-grained 2024 aluminium alloy. Sci Technol Weld Join 2017;22:110−19. doi:10.1080/13621718.2016.1203177; Milan MT, Bose WW, Tarpani JR. Fatigue crack growth behavior of friction stir welded 2024-T3 aluminum alloy tested under accelerated salt fog exposure. Mater Perform Charact 2014;3:20130036. doi:10.1520/MPC20130036, [15,16].

As was discussed in the previous section, the FCG in FSW response is dependent upon many factors. The environment in which a fatigue crack grows will strongly affect the materials response, as shown in Fig. 4.8. One must always use Paris constants that are particular to the environmental conditions, as the FCGR can be significantly different in air as compared to salt conditions (salt fog and submerged conditions) as example. The effect of environment on fatigue in FSW is discussed further in Chapter 6, Extreme Conditions and Environments.

4.2.2 5XXX Series Aluminum
FCG data for 5xxx series aluminum alloys are provided in Fig. 4.9. In particular, the effects of weld region, weld process parameters, and load ratio are highlighted.

4.2.3 6xxx-Series Aluminum
Fig. 4.10 shows examples for FCG data for AA6082 for both BMs and FSW. FCGRs are higher for AA6082 FSW compared to BM. In contrast, the FCGRs were observed to be similar at higher stress

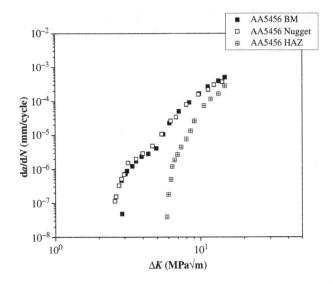

Figure 4.9 FCG data for AA5456 BM, Nugget (weld), and HAZ. Data replotted from Fonda RW, Pao PS, Jones HN, Feng CR, Connolly BJ, Davenport AJ. Microstructure, mechanical properties, and corrosion of friction stir welded Al 5456. Mater Sci Eng A 2009;519:1–8. doi:10.1016/j.msea.2009.04.034; Schwinn J, Besel M, Alfaro Mercado U. Experimental determination of accurate fatigue crack growth data in tailored welded blanks. Eng Fract Mech 2016;163:141–59. doi:10.1016/j.engfracmech.2016.07.006, [17,18].

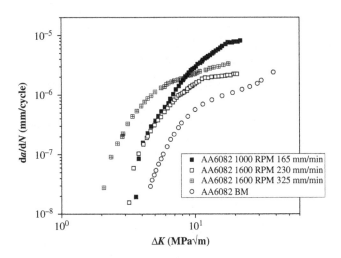

Figure 4.10 FCG data for AA6082 BM and FSW as a function of process parameters. Data replotted from Cavaliere P, Santis AD, Panella F, Squillace A. Thermoelasticity and CCD analysis of crack propagation in AA6082 friction stir welded joints. Int J Fatigue 2009;31:385–92. doi:10.1016/j.ijfatigue.2008.07.016, [19].

intensities for AA6013 (Fig. 4.4). However, near threshold, the FCGRs were significantly higher for FSW.

4.2.4 7xxx-Series Aluminum

FCG data for 7xxx-series aluminums are provided in Fig. 4.11. One will note that for the AA7075 data provided, the alloys FCG response is somewhat insensitive to weld region and post-weld peening.

4.2.5 Magnesium, Ferritic Stainless Steel, and Titanium Alloys

FCG data collected from literature on magnesium alloys (Fig. 4.12A), a ferritic stainless steel (Fig. 4.12B), and Ti6Al4 (Fig. 4.13) are

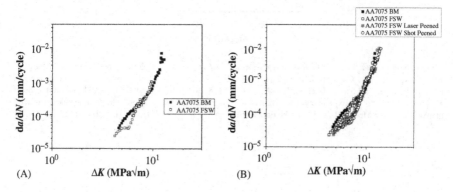

Figure 4.11 (A) FCG data for AA7075 BM and FSW. (B) FCG data for AA7075 BM, as welded FSW, shot peened FSW, and laser peened FSW. Data replotted from Hatamleh O, Forth S, Reynolds AP. Fatigue crack growth of peened friction stir-welded 7075 aluminum alloy under different load ratios. J Mater Eng Perform 2010;19:99–106. doi:10.1007/s11665-009-9439-1, [20].

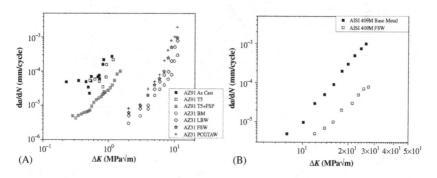

Figure 4.12 FCG data for (A) magnesium alloy. Data replotted from [21,22]. FCG data (B) ferritic stainless steel. Data replotted from Lakshminarayanan AK, Balasubramanian V. Assessment of fatigue life and crack growth resistance of friction stir welded AISI 409M ferritic stainless steel joints. Mater Sci Eng A 2012;539:143–53. doi:10.1016/j.msea.2012.01.071, [23].

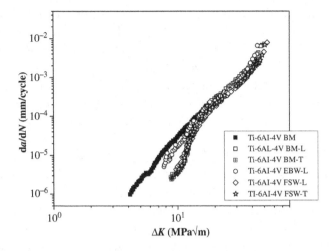

Figure 4.13 FCG data for titanium alloy Ti-6Al-4V. Data replotted from Pasta S, Reynolds AP. Residual stress effects on fatigue crack growth in a Ti-6Al-4V friction stir weld. Fatigue Fract Eng Mater Struct 2008;31:569−80. doi:10.1111/j.1460-2695.2008.01258.x.; Edwards PD, Ramulu M. Comparative study of fatigue and fracture in friction stir and electron beam welds of 24 mm thick titanium alloy Ti-6Al-4 V. Fatigue Fract Eng Mater Struct 2016;39:1226−40. doi:10.1111/ffe.12434, [24,25].

provided in this section. Note that while the change in BM between the AZ91 and AZ31 magnesium alloys has a drastic effect on the FCG response, the FSW process will also affect the response (Fig. 4.11A). The FSW process was observed to retard FCG in the AZ91 alloy but increased FCG in the AZ31 alloy. Finally, based on the data presented in Fig. 4.13, FCG of Ti6Al4 appears to be insensitive to the FSW process.

4.3 EFFECT OF RESIDUAL STRESS ON FCG

Although residual stress introduced during FSW is generally much lower compared to other welding techniques, studies have shown that FCG can still be affected by the relatively low residual stresses that develop during the FSW process [1]. Although it is possible to study the effect of residual stresses in overlap FSW joints [26], most of the studies on the impact of residual stresses on fatigue are focused on butt joints. Work on studying the impact of residual stresses on fatigue performance or FCG in FSW joints is somewhat limited due the complex nature of quantifying the residual stress within the weld and subsequent FCG testing.

A survey of literature shows that in some cases, the as-welded FSW joint exhibited improved FCG resistance compared to the BM

[11,24,27], whereas in contrast other studies reported similar or lower FCG resistance compared to the BM [7,14]. In particular, Hong et al. [28], observed that the base metal exhibited higher FCGRs compared to the dynamically recrystallized weld zone of the FSW. They concluded that the beneficial compressive residual stresses from the FSW process reduced the stress intensity ahead of the crack tip along with a favorable microstructure to greatly reduce the fatigue FCG of FSW joint. In another study, Sowards et al. [27] characterized the fatigue crack propagation in FSW microalloyed pipeline steels X-80 and found that the base metal generally exhibited higher FCGRs compared to FSW joints due to similar mechanisms to the study by Hong et al. In another study, Fratini et al. [29] observed that in FSW of 2024 aluminum alloy, FCGRs were most affected outside the weld zone, but once the fatigue crack propagates in the weld zone, the crack growth rate is more influenced by the microstructure and hardness changes. This observation suggests that complex compression and tensile residual stresses present in FSW can have a strong impact on the fatigue behavior. Additionally, Ma et al. [11] studied the effect of specimen size on residual stress and FCGR in FSW 2195-T8 alloy. In compact test specimens, the FCGRs were 5−10 times slower compared to the base metal, but this difference was more prominent in early fatigue crack initiation stage. However, when the fatigue cracks grew longer, the difference in the FCGRs of the FSW and BM minimized, as shown in the Fig. 4.14 [11].

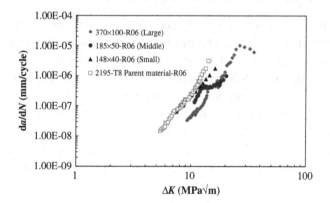

Figure 4.14 Crack growth rate da/dN versus ΔK in three different sample sizes; compact tension samples crack plane parallel to the weld line [11].

The observation from studies on residual stress in FSW indicates that there is no single variable that controls the magnitude of residual stresses in an FSW joint. As a design engineer, it is important to realize that like any other manufacturing and welding process, FSW has the potential to introduce residual stresses in the weld joint which will impact its mechanical strength and particularly the fatigue performance. Therefore, care must be taken to avoid or mitigate the introduction of excessive residual stresses particularly, tensile residual stresses in the weld joint which can negatively impact the fatigue life. In mass produced critical FSW structures, the measurement of residual stress may be costly; therefore, it is advised that in the early stages of design, engineers and researchers should make use of nondestructive methods such as neutron diffraction to quantitatively measure the residual stresses in laboratory-developed FSW joints. In addition, comprehensive laboratory fatigue testing should be followed to quantify the effect of residual stresses on fatigue properties to understand what process parameters or variable influence the residual stresses for their given material and joint type. For example, if residual stresses are introduced due to excessive clamping forces, then corrective measures should be applied to help offset these forces. In case of residual stresses purely introduced due to geometry of the joint or material-centric properties or due to welding process variables, then additional secondary processes such as heat treatment, peening, or mechanical stress relief should be performed to relieve the tensile residual stresses or introduce beneficial compressive residual stresses in the FSW joint.

Regarding techniques to reduce residual stresses, Dorman et al. [30] observed in FSW 2XXX aluminum alloys, laser peening post-FSW introduced beneficial compressive residual stresses in the weld joint that reduced the FCGR. Mechanical stress relief post-FSW have also been used to reduce the transverse and longitudinal residual stresses in 2XXX aluminum alloys [29,31]. In addition, a transient thermal tensioning (TTT) method in which a moving heat source located on both sides of the weld line has also been studied to relieve residual stresses during the FSW [32]. The TTT treatment of FSW was shown to have a positive effect on improving the fatigue resistance in FSW 2xxx aluminum alloys. Some of the methods discussed in this section are just a few examples which have been used to reduce the effect of residual stresses on FCG in FSW joints.

4.4 CRACK GROWTH MECHANISMS IN FRICTION STIR SPOT WELDS

While this chapter focuses mainly on FCG concepts and associated behaviors based on data generated mainly from CT tests of FSBW, understanding FCG mechanisms in lap-shear FSSW joints is also of interest. Because the lap-shear FSSW coupon naturally shields direct planar view of FCG within the joint, characterization and associated understanding of the mechanisms of FCG in spot welds is difficult. As such, in this section, we briefly present a previously unpublished study on FCG of FSSW in magnesium alloys. The following describes a novel fatigue test method for quantifying FCG in FSSW joints and the corresponding results of these unique experiments.

Due to the geometrical nature of the coupons, a specialized experimental setup was employed to allow for direct imaging of the FCG in FSSW coupons. The experimental process involved conducting interrupted fatigue tests based on a unique specimen design that elucidated the fatigue crack incubation and propagation mechanisms within the interior of the weld nugget of the lap-shear coupon. The FSSW joints were fabricated from 2 mm thick AZ31 Mg alloy sheets. Information on the preparation of the FSSW coupons used in this study is detailed elsewhere [5,33].

In order to image directly and observe fatigue incubation and crack growth, the FSSW coupons were sectioned through the center line of the weld so as to reveal the internal weld nugget of the FSSW coupon. The exposed weld nugget was then polished using standard metallographic methods. Due to the now unsymmetrical bending of the specimen under mechanical loading, a pair of special brackets was employed to provide the constraint that was removed when the coupon was sectioned. Fig. 4.15 shows the half-width FSSW coupon and the bracket inserted in the MTS load frame. The interrupted fatigue tests were carried out at $R = 0$, and at room temperature and relativity humidity. The fatigue tests on the half-width coupons were interrupted at intervals of 100 cycles, and then the weld zone and adjacent faying surface were imaged under an optical microscope.

As noted in previous chapters, FSSW joints typically contain a faying surface that often times exhibits a hook-like shape as shown in Fig. 4.16. This hooking of the faying surface is a result of trapped

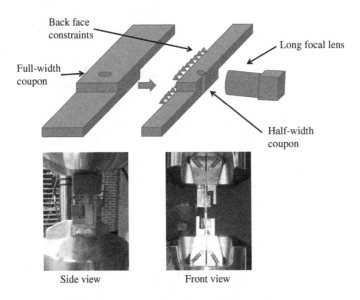

Figure 4.15 Illustration of half-width coupon setup for crack growth fatigue testing.

oxide films that are displaced upward toward the tool due to the plastic flow of the material resulting from the downward plunge of the pin into the bottom sheet. In fact, the degree of oxide distribution at the faying surface is greatly influenced by the tool geometry and tool rotation speed [5]. Although this type of hooking feature in FSSW is well established in literature, we present an image of it here to aid in the discussion of the FCG experiments present later in this section.

Normally, under monotonic static tensile loading, FSSW coupons can fail by nugget pullout. This failure mode is consistent with this range of welding parameters [33] on AZ31 Mg alloy joined in an overlap configuration. Nugget pullout failure under monotonic loading is facilitated by the high shear and tensile stress exerted on the interfacial hook. In fact, it is this shear loading that initiates the crack, and as the crack propagates through the nugget, the crack mode translates to a tensile load leading to final failure. Under cyclic loading, however, the failure mode is significantly different. The failure mode under cyclic loading is dependent on the fraction of cyclic load relative to the ultimate strength of the joint. At a higher percentage of the ultimate strength, the final fracture is similar to tensile lap-shear testing resulting in nugget pullout. However, at a lower fraction of the ultimate strength, final failure occurs by top sheet failure. The top sheet failure

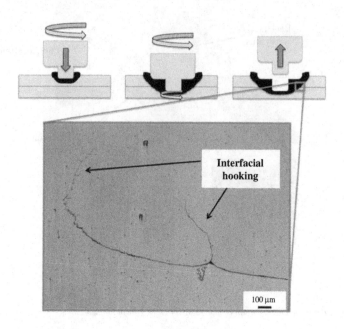

Figure 4.16 A schematic of the FSSW process and a magnified view of the interfacial hooking within the FSSW nugget.

is a result of FCG from the interfacial hook leading to continued FCG through the top sheet.

Fig. 4.17 shows a representative collection of optical microscopic images taken during the interrupted fatigue experiments of the half-width coupon. From Fig. 4.17, the interfacial hook is clearly observed prior to loading. For brevity's sake, only an image taken every 200th cycle is shown. The result of the interrupted fatigue tests on the FSSW show that the crack incubation occurs from the hooking feature within 10% of the total cycles to failure. Furthermore, the results show that the fatigue crack grew in mode I to the free surface of the FSSW sample. These observations are significant because these experiments are the first to characterize the FCG in FSSW of any type of material. While this chapter is focused on FCG, it is important to address that the results presented here confirm the assumption that the fatigue crack incubation stage represents a relatively small percentage of total number of cycles to failure in FSSW lap-joints. In fact, this confirmation is important because many fatigue modeling approaches assume the number of cycles to incubate a fatigue crack in FSSW joints is

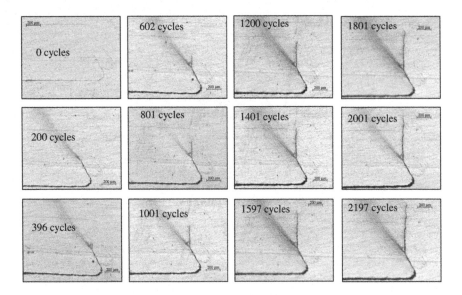

Figure 4.17 Fatigue crack growth in friction stir spot welded coupons in half-width lap-shear configuration subjected to an equivalent maximum cyclic load of 2 kN and a load ratio, R = 0.

negligible. However, we acknowledge that the experiments performed in this study were not performed on coupons subjected to loads that produced number of cycles to failure in the 10^6 cycles or higher. Thus, it is likely that the number of cycles needed to incubate a dominate fatigue crack could represent a significant portion of the total number of cycles to failure. However, more experiments are needed to confirm incubation mechanisms in FSSW; thus, this interrupted test method could be used to explore the high-cycle fatigue regime.

Fig. 4.18 shows the FCGR plotted against the fraction of total number of cycles calculated directly from the images shown in Fig. 4.17. Also included in Fig. 4.18 is the experimental tests results of a second similar interrupted fatigue test. The significance of Fig. 4.18 is that despite the observed crack growth scatter normally seen in short FCG experiments [34], the average FCGR remains constant. As such, these results add credibility to the assumption of constant FCG that has been used in modeling the sub-surface fatigue behavior in FSSW joints.

In summary, a novel testing approach was developed to directly measure FCG in FSSW. Interrupted high-resolution imaging using optical microscope was allowed for high fidelity measurements.

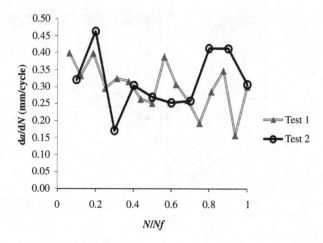

Figure 4.18 Fatigue crack growth rate versus the fraction of the total number of cycles for friction stir spot welded coupons in half-width lap-shear configuration subjected to an equivalent maximum cyclic load of 2 kN and a load ratio, R = 0.

The experimental results revealed that the fatigue cracks grew at a constant FCGR and initiated at the tip of the interfacial hook. The final conclusions of the experimental results suggest that the FCG in FSSW is not strongly dependent on the microstructure of the stir zone but rather is largely dependent on the geometrical features of the interfacial hooking which tends to favor a fracture mechanics modeling approach.

REFERENCES

[1] John R, Jata K, Sadananda K. Residual stress effects on near-threshold fatigue crack growth in friction stir welds in aerospace alloys. Int J Fatigue 2003;25:939−48. Available from: https://doi.org/10.1016/j.ijfatigue.2003.08.002.

[2] Edwards P, Ramulu M. Fracture toughness and fatigue crack growth in Ti-6Al-4V friction stir welds. Fatigue Fract Eng Mater Struct 2015;38:970−82. Available from: https://doi.org/10.1111/ffe.12291.

[3] Biro AL, Chenelle BF, Lados DA. Processing, microstructure, and residual stress effects on strength and fatigue crack growth properties in friction stir welding: a review. Metall Mater Trans B 2012;43:1622−37. Available from: https://doi.org/10.1007/s11663-012-9716-5.

[4] Jordon JB, Horstemeyer MF, Daniewicz SR, Badarinarayan H, Grantham J, Jordon JB, et al. Fatigue characterization and modeling of friction stir spot welds in magnesium AZ31 alloy. J Eng Mater Technol 2010;132:041008. Available from: https://doi.org/10.1115/1.4002330.

[5] Rao HM, Jordon JB, Barkey ME, Guo YB, Su X, Badarinarayan H. Influence of structural integrity on fatigue behavior of friction stir spot welded AZ31 Mg alloy. Mater Sci Eng A 2013;564:369−80. Available from: https://doi.org/10.1016/j.msea.2012.11.076.

[6] Moraes JFCFC, Rodriguez RII, Jordon JBB, Su X. Effect of overlap orientation on fatigue behavior in friction stir linear welds of magnesium alloy sheets. Int J Fatigue 2017;100:1−11. Available from: https://doi.org/10.1016/j.ijfatigue.2017.02.018.

[7] Biallas G. Effect of welding residual stresses on fatigue crack growth thresholds. Int J Fatigue 2013;50:10−17. Available from: https://doi.org/10.1016/j.ijfatigue.2012.07.002.

[8] Moreira PMGP, de Jesus AMP, de Figueiredo MAV, Windisch M, Sinnema G, de Castro PMST. Fatigue and fracture behaviour of friction stir welded aluminium-lithium 2195. Theor Appl Fract Mech 2012;60:1−9. Available from: https://doi.org/10.1016/j. tafmec.2012.06.001.

[9] Cavaliere P, De Santis A, Panella F, Squillace A. Effect of anisotropy on fatigue properties of 2198 Al-Li plates joined by friction stir welding. Eng Fail Anal 2009;16:1856−65. Available from: https://doi.org/10.1016/j.engfailanal.2008.09.024.

[10] Okada T, Machida S, Nakamura T, Tanaka H, Kuwayama K, Asakawa M. Fatigue crack growth of friction-stir-welded aluminum alloy. J Aircr 2017;54:737−46. Available from: https://doi.org/10.2514/1.C034119.

[11] Ma YE, Staron P, Fischer T, Irving PE. Size effects on residual stress and fatigue crack growth in friction stir welded 2195-T8 aluminium—Part I: Experiments. Int J Fatigue 2011;33:1417−25. Available from: https://doi.org/10.1016/j.ijfatigue.2011.05.006.

[12] Ma YE, Zhao ZQ, Liu BQ, Li WY. Mechanical properties and fatigue crack growth rates in friction stir welded nugget of 2198-T8 Al-Li alloy joints. Mater Sci Eng A 2013;569:41−7. Available from: https://doi.org/10.1016/j.msea.2013.01.044.

[13] Ma YE, Xia ZC, Jiang RR, Li WY. Effect of welding parameters on mechanical and fatigue properties of friction stir welded 2198 T8 aluminum−lithium alloy joints. Eng Fract Mech 2013;114:1−11. Available from: https://doi.org/10.1016/j.engfracmech.2013.10.010.

[14] Hatamleh O. A comprehensive investigation on the effects of laser and shot peening on fatigue crack growth in friction stir welded AA 2195 joints. Int J Fatigue 2009;31:974−88. Available from: https://doi.org/10.1016/j.ijfatigue.2008.03.029.

[15] Wang W, Qiao K, Wu JL, Li TQ, Cai J, Wang KS. Fatigue properties of friction stir welded joint of ultrafine-grained 2024 aluminium alloy. Sci Technol Weld Join 2017;22:110−19. Available from: https://doi.org/10.1080/13621718.2016.1203177.

[16] Milan MT, Bose WW, Tarpani JR. Fatigue crack growth behavior of friction stir welded 2024-T3 aluminum alloy tested under accelerated salt fog exposure. Mater Perform Charact 2014;3:20130036. Available from: https://doi.org/10.1520/MPC20130036.

[17] Fonda RW, Pao PS, Jones HN, Feng CR, Connolly BJ, Davenport AJ. Microstructure, mechanical properties, and corrosion of friction stir welded Al 5456. Mater Sci Eng A 2009;519:1−8. Available from: https://doi.org/10.1016/j.msea.2009.04.034.

[18] Schwinn J, Besel M, Alfaro Mercado U. Experimental determination of accurate fatigue crack growth data in tailored welded blanks. Eng Fract Mech 2016;163:141−59. Available from: https://doi.org/10.1016/j.engfracmech.2016.07.006.

[19] Cavaliere P, Santis AD, Panella F, Squillace A. Thermoelasticity and CCD analysis of crack propagation in AA6082 friction stir welded joints. Int J Fatigue 2009;31:385−92. Available from: https://doi.org/10.1016/j.ijfatigue.2008.07.016.

[20] Hatamleh O, Forth S, Reynolds AP. Fatigue crack growth of peened friction stir-welded 7075 aluminum alloy under different load ratios. J Mater Eng Perform 2010;19:99−106. Available from: https://doi.org/10.1007/s11665-009-9439-1.

[21] Uematsu Y, Tokaji K, Fujiwara K, Tozaki Y, Shibata H. Fatigue behaviour of cast magnesium alloy AZ91 microstructurally modified by friction stir processing. Fatigue Fract Eng Mater Struct 2009;32:541−51. Available from: https://doi.org/10.1111/j.1460-2695.2009.01358.x.

[22] Padmanaban G, Balasubramanian V, Reddy GM. Fatigue crack growth behaviour of pulsed current gas tungsten arc, friction stir and laser beam welded AZ31B magnesium alloy joints. J Mater Process Technol 2011;211:1224−33. Available from: https://doi.org/10.1016/j. jmatprotec.2011.02.003.

[23] Lakshminarayanan AK, Balasubramanian V. Assessment of fatigue life and crack growth resistance of friction stir welded AISI 409M ferritic stainless steel joints. Mater Sci Eng A 2012;539:143–53. Available from: https://doi.org/10.1016/j.msea.2012.01.071.

[24] Pasta S, Reynolds AP. Residual stress effects on fatigue crack growth in a Ti-6Al-4V friction stir weld. Fatigue Fract Eng Mater Struct 2008;31:569–80. Available from: https://doi.org/10.1111/j.1460-2695.2008.01258.x.

[25] Edwards PD, Ramulu M. Comparative study of fatigue and fracture in friction stir and electron beam welds of 24 mm thick titanium alloy Ti-6Al-4 V. Fatigue Fract Eng Mater Struct 2016;39:1226–40. Available from: https://doi.org/10.1111/ffe.12434.

[26] Solanki KN, Jordon JB, Whittington W, Rao H, Hubbard CR. Structure–property relationships and residual stress quantification of a friction stir spot welded magnesium alloy. Scr Mater 2012;66:797–800. Available from: https://doi.org/10.1016/j.scriptamat.2012.02.011.

[27] Sowards JW, Gnäupel-Herold T, David McColskey J, Pereira VF, Ramirez AJ. Characterization of mechanical properties, fatigue-crack propagation, and residual stresses in a microalloyed pipeline-steel friction-stir weld. Adv. Nondestruct. Eval. II 2015;vol. 1:318–23. Available from: https://doi.org/10.1016/j.matdes.2015.09.049 Elsevier B.V.

[28] Hong S, Kim S, Lee CG, Kim SJ. Fatigue crack propagation behavior of friction stir welded Al-Mg-Si alloy. Scr Mater 2006;55:1007–10. Available from: https://doi.org/10.1016/j.scriptamat.2006.08.012.

[29] Fratini L, Pasta S, Reynolds AP. Fatigue crack growth in 2024-T351 friction stir welded joints: Longitudinal residual stress and microstructural effects. Int J Fatigue 2009;31:495–500. Available from: https://doi.org/10.1016/j.ijfatigue.2008.05.004.

[30] Dorman M, Toparli MB, Smyth N, Cini A, Fitzpatrick ME, Irving PE. Effect of laser shock peening on residual stress and fatigue life of clad 2024 aluminium sheet containing scribe defects. Mater Sci Eng A 2012;548:142–51. Available from: https://doi.org/10.1016/j.msea.2012.04.002.

[31] Bussu G, Irving PE. The role of residual stress and heat affected zone properties on fatigue crack propagation in friction stir welded 2024-T351 aluminium joints. Int J Fatigue 2002;25:77–88. Available from: https://doi.org/10.1016/S0142-1123(02)00038-5.

[32] Ilman MN, Kusmono, Iswanto PT. Fatigue crack growth rate behaviour of friction-stir aluminium alloy AA2024-T3 welds under transient thermal tensioning. Mater Des 2013;50:235–43. Available from: https://doi.org/10.1016/j.matdes.2013.02.081.

[33] Solanki KN, Jordon JB, Whittington W, Rao H, Hubbard CR. Structure-property relationships and residual stress quantification of a friction stir spot welded magnesium alloy. Scr Mater 2012;66. Available from: https://doi.org/10.1016/j.scriptamat.2012.02.011.

[34] Jordon JB, Bernard JD, Newman Jr. JC. Quantifying microstructurally small fatigue crack growth in an aluminum alloy using a silicon-rubber replica method. Int J Fatigue 2012;36:206–10. Available from: https://doi.org/10.1016/j.ijfatigue.2011.07.016.

CHAPTER 5

Fatigue Modeling of Friction Stir Welding

5.1 INTRODUCTION

Designing against fatigue in friction stir welding (FSW) can be a complex and difficult endeavor. Similar to conventional fusion welds, accurately predicting the number of cycles to failure depends largely on one's knowledge of the structural integrity of the weld and thorough understanding of the boundary conditions, that is, loading configuration, expected service loads, environment, etc. Although international standards [1,2] for conventional fusion welding provide guidelines for the design, the recommendations for FSW are still being developed. From a practical perspective, modeling the fatigue life of FSW joints largely depends on the structure type, industry culture, the design criteria, and maintenance strategy. For conventional welds in aluminum alloys, Maddox [3] summarized the commonly used fatigue life assessment approaches: stress-life ($S-N$) curve for nominal stresses; $S-N$ curve with the use of notch analysis; $S-N$ curve with use of structural stresses; and fracture mechanics. These broad categories also represent the methods used in modeling fatigue in FSW. As such, this chapter presents brief overviews of approaches for fatigue life assessment in FSW. Similar to conventional fusion welding, this chapter will cover the safe-life approaches adapted for FSW along with examples of stress and strain-based concepts. In addition, this chapter includes an overview of the structural stress approach favored in some industries due to the computational efficiency and ease of method. Again, similar to approaches used in fatigue of base materials and other welded joints, the damage tolerance approach has been shown as a valid method for predicting the fatigue behavior in FSW based on crack growth concepts. Several examples demonstrating the use of linear elastic fracture mechanics (LEFM) are shown in order to aid the reader. Lastly, due to the graded microstructure and nature of defects, predicting fatigue in FSW can lend itself to microstructure-sensitive modeling methods that combine micromechanical simulations and fracture mechanics

Fatigue in Friction Stir Welding. DOI: https://doi.org/10.1016/B978-0-12-816131-9.00005-2
© 2019 Elsevier Inc. All rights reserved.

approaches to capture the complex multistage mechanisms of fatigue crack incubation and growth.

5.2 STRESS-LIFE APPROACH

In general, the $S-N$ approach for assessing fatigue behavior is essentially based on the use of fatigue life curves for specific welds in conjunction with nominal stresses [3]. Thus, from a design perspective, the number of cycles to failure is determined by the associated stress from the $S-N$ curve. The benefit of this approach is that local stresses like notch root stresses and other geometrical features like the surface finish and weld defects in FSW do not need to be explicitly modeled because these stresses are inherently a function of the $S-N$ data for a specific weld. However, one must be careful in the use of the $S-N$ approach because the fatigue behavior of various FSW joint types and loading configurations can have significantly different fatigue lives. As such, it is imperative that the choice of the laboratory fatigue coupon be highly representative of the actual FSW joint and loading conditions in order to prevent nonconservative estimations of fatigue lives.

$S-N$ fatigue testing was traditionally carried out under rotating bending loading, but uniaxial testing in a hydraulic load frame is more common nowadays. The stress-life tests are usually conducted under fully revered ($R = -1$) loading or some positive ratio ($R > 0$). A sampling of literature on $S-N$ curves of friction stir butt welding (FSBW) reveals that most studies focus on aluminum alloys [4–27]. Other studies report on the FSBW of magnesium alloys [28–30], steels [31,32] titanium alloys [33], and dissimilar metals [34]. Although friction stir linear welding (FSLW) and FSSW lap-joints are also usually conducted in load control, typical modeling of the lap-joints are discussed later in this chapter. An example of the $S-N$ curve for FSBW of aluminum alloys taken from numerous studies is presented in Fig. 5.1 [35].

The $S-N$ curves used in design codes are usually derived from linear regression analysis of the log–log plot of the experimental fatigue results [3]. In addition, the design curves are usually based on a statistical lower bound, where this lower bound can be set as low as two standard deviations below the mean [3]. The $S-N$ curve is represented by the expression

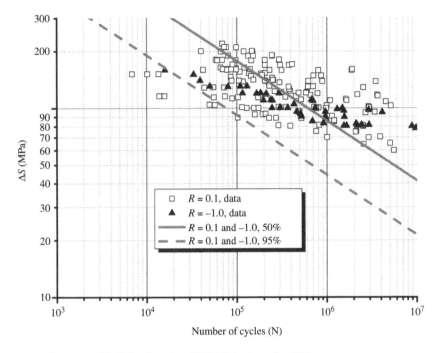

Figure 5.1 Comparison of S−N data for various FSBW of aluminum alloys [35].

$$S^m N = A \qquad (5.1)$$

where m and A are constants derived from the regression analysis. Note, that for most $S-N$ curves, as the fatigue strength increases, the slope of the curve also increases. Although design codes like the Eurocode 9 [1] and International Institute of Welding (IIW) [2] do not specify a type of specific welding process, sound FSW joints generally meet or exceed the recommended $S-N$ performance requirements [35–39]. Fig. 5.2 [35] shows an example of the 95% survival probability $S-N$ curve representing numerous FSBW data sets derived from Fig. 5.1 along with the design curves from the IIW. Notice that the 95% survival probability $S-N$ curve meets or exceeds a majority of IIW design curves (FAT12, FAT25, and FAT36). The slight subperformance of the FSW at 95% survival probability compared to the FAT36 curve is likely due to the lack of quality control of the data set. Nonetheless, FSW are generally considered to perform better than the design curves for conventional welds. However, future work is needed to establish design curves specifically for FSW.

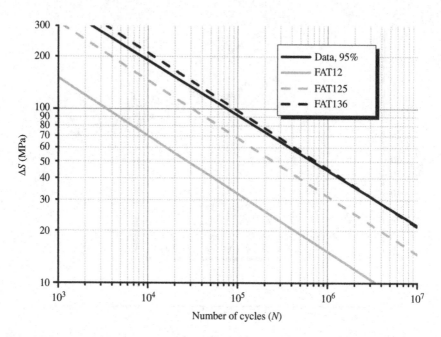

Figure 5.2 Comparison of the experimental 95% survival probability S−N curve and design curves [35].

From an engineering/designing perspective, the use of the $S-N$ curve also requires understanding the loading conditions. Although some FSW joints will experience primarily constant amplitude loading, others will be subjected to variable amplitude. As such, accumulative damage rules are commonly used to equate some variable amplitude load case to $S-N$ curves generated under constant amplitude. More details regarding variable amplitude are provided in Chapter 6, Extreme Conditions and Environments.

The drawback to employing the $S-N$ curve in conjunction with a nominal stress design analysis is the restriction that the $S-N$ curve only applies to that specific joint and loading configuration. An $S-N$ curve developed for an FSW joint subjected to lap-shear loading would not be applicable for the same FSW joint subjected to coach-peel loading. A new $S-N$ curve would need to be developed for the coach-peel configuration. Fig. 5.3 shows an example of an $S-N$ curve for FSW T-Joints subjected to different loading conditions and load ratios. Notice in Fig. 5.3 the significant difference in the stress ranges for a corresponding number of cycles. It is important to note that these two different data sets were tested at different load ratios and that,

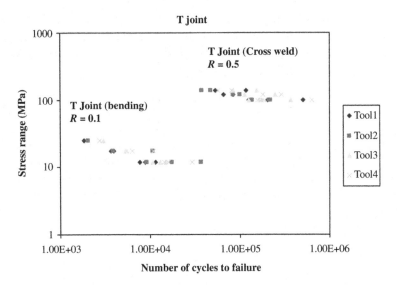

Figure 5.3 Example of a stress-life (S–N) curve for friction stir welding (FSW) lap-shear and T-joint coupons [12].

although likely not causing a significant difference, there could be some effect due to the load ratio. In an attempt to address the problem of needing separate $S-N$ curves, a few studies [12,16,40,41] have used notch stress analysis to explicitly capture the differences in local stress concentrations and applied loading between different $S-N$ curves with some degree of success.

5.3 STRAIN-LIFE APPROACH

Although many structural service loads are below the yield of the material or the weld, plastic deformation from cyclic loading can occur, especially in areas where stress concentrations exist. As such, the strain-life approach is considered to be more appropriate for modeling fatigue under high loads where yielding will occur compared to the stress-life approach. The assumption of the strain-life approach is that $Kt = 1$ specimens (stress concentration equal to one) tested under strain-control fatigue can model the damage from fatigue at the notch root [42]. There are few strain-controlled fatigue studies on FSBW on various metals including aluminum alloys [43–49], titanium alloy [50], and magnesium alloys [51,52].

In the strain-life approach, the low-cycle and high-cycle fatigue properties can be estimated by resolving the elastic and plastic strain amplitude components from the total strain amplitude. As such, the total strain amplitude can be expressed as

$$\frac{\Delta\varepsilon}{2} = \frac{\Delta\epsilon_e}{2} + \frac{\Delta\epsilon_p}{2} \tag{5.2}$$

where, $\frac{\Delta\varepsilon}{2}$ is the total strain amplitude, $\frac{\Delta\epsilon_e}{2}$ is the elastic strain amplitude component, and $\frac{\Delta\epsilon_p}{2}$ is the plastic strain amplitude component. The elastic and plastic strain amplitude components can be further expressed by using the Basquin (Eq. 5.3) and Coffin–Manson (Eq. 5.4) equations, which are defined as

$$\frac{\Delta\epsilon_e}{2} = \frac{\sigma_f'}{E}(2N_f)^b \tag{5.3}$$

$$\frac{\Delta\epsilon_p}{2} = \epsilon_f'(2N_f)^c \tag{5.4}$$

where E is the elastic modulus, σ_f' is the fatigue strength coefficient, b is the fatigue strength exponent, ϵ_f' is the fatigue ductility coefficient, c is the fatigue ductility exponent, and N_f is the number of cycles to failure.

The values of the coefficients and exponents in Eqs. 5.3 and 5.4 are determined by performing linear regression on each equation separately. Although it might not be particularly obvious, the strain measurement device used in strain-control fatigue testing, which is typically an extensometer, only measures total strain and does not distinguish between elastic and plastic components. It is important to note that the calibration of the strain-life parameters is based on stabilized hysteresis loops. In some cases, the hysteresis loops never stabilize and thus it is common to use the hysteresis loop from the midway point in the number of cycles to failure for a particular specimen, that is, half-life cycle. The stabilized hysteresis loops can then be used to determine the elastic and plastic fraction of the strain amplitudes for each data point, where the strain-life coefficients and exponents can be determined. For more details regarding the strain-life method, the reader is referred to additional resources on fatigue of metals [42,53]. Fig. 5.4A shows an example of the strain-life fit of the AA2099 base material, Fig. 5.4B and C show the strain-life fits for FSW of AA2099 for two different welding parameters. Note that the elastic and plastic

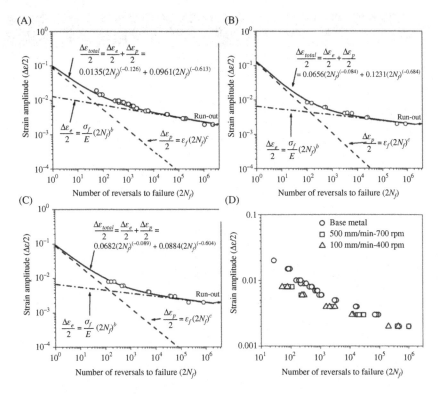

Figure 5.4 Strain-life fatigue behavior of (A) Base metal AA2099, (B) FSW AA2099 with welding Parameters I (400 rpm/100 mm/min), (C) FSW AA2099 with welding Parameters II (700 rpm/500 mm/min), and (D) comparison between the fatigue lives of the base metal and two welding parameters [43]. Reprinted with permission from author.

components are plotted separately in addition to the total strain-life equation and exhibit a linear stress–strain relationship when plotted in a log–log format. Lastly, Fig. 5.4D shows a comparison of the base material and FSBW experimental results. For reference purposes, Table 5.1 presents representative strain-life parameters for various FSBW of aluminum, magnesium alloys, and titanium alloys.

It is important to briefly note that the strain-life approach can be modified to account for mean stresses that may develop from non-fully reversed loading. Similar to base materials, the Smith, Watson, and Topper (SWT) equation can be used for FSW and takes on the form:

$$\sigma_{max} \frac{\Delta\varepsilon}{2} = \frac{\left(\sigma_f'\right)^2}{E}\left(2N_f\right)^{2b} + \sigma_f'\epsilon_f'\left(2N_f\right)^{b+c} \qquad (5.5)$$

Table 5.1 Representative Strain-life Parameters for FSBW of Aluminum, Magnesium, and Titanium Alloys						
Alloy	K' (MPa)	n'	σ_f' (MPa)	b	ε_f'	c
2219 Al Alloy [45]	248	0.10	517	− 0.09	0.64	− 0.79
6061 Al Alloy [48]	520	0.11	509	− 0.09	0.29	− 0.71
7050 Al Alloy [49]	808	0.126	1477	− 0.159	0.026	− 0.63
AZ31 Mg Alloy [52]	467	0.21	805	− 0.22	0.24	− 0.66
AZ91 Mg Alloy [51]	650	0.21	549	− 0.16	0.081	− 0.58
Ti-6Al-4/Ti17 [50]	2968	0.20	2916	− 0.18	0.70	− 0.88

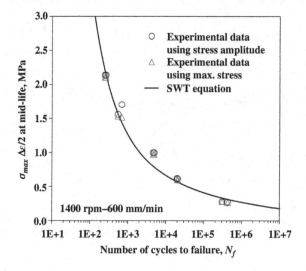

Figure 5.5 Comparison of the Smith, Watson, and Topper Strain-life equation for FSW AA6061 [48]. Reprinted with permission from Springer Nature.

where σ_{max} is the maximum stress at the stabilized hysteresis or half-life cycle. Several studies [48,54] have reported good agreement between the SWT and experimental FSW fatigue data as depicted in Fig. 5.5. However, further strain-life testing under a range of load ratios is suggested to fully assess the appropriateness of the SWT modification for FSW applications.

In the event of service loads acting on the FSW that are not constant amplitude, the use of accumulative damage rules is suggested. We again point the reader to Chapter 6, Extreme Conditions and Environments, where the application of the accumulative damage models such as the Miner's rule is presented and discussed.

Similar to the use of notch analysis in the $S-N$ approach, notch analysis with the strain-life approach has been used on FSLW lap-joints with some limited results. Wang et al. [55] reported that nonconservative results were achieved in the high-cycle regime. In addition, the strain-life approach produced a good deal of scatter by not collapsing the curves of the different materials. As such, Wang et al. suggested that the inaccuracy of the strain-life approach in the high-cycle regime and the undesirable scatter of the results were due to lack of information regarding the notch geometry incorporated into the finite element analysis (FEA) and inappropriate strain-life parameters. However, in the same study, Kang et al. showed that the structural stress approach, which is discussed in Section 5.4, produced more accurate results and was less dependent on additional parameters that can be difficult to determine (strain-life parameters). In a similar study [15], difficulty in selecting appropriate strain-life parameters resulted in significant discrepancy in predicting the low-cycle fatigue regime of the notched FSBW samples.

5.4 STRUCTURAL STRESS APPROACH

The main problem associated with the notch analysis is the need for a highly refined mesh. As such, achieving accurate results using notch analysis is that the results become mesh size dependent. This also can lead to large computational costs, especially if large structures with multiple welds are analyzed. Thus, to overcome the problem of needing to perform highly mesh-dependent finite element analysis on details of the weld, one could determine the structural stresses around the FSW joint with a simple and relatively coarse mesh. In order to determine the structural stresses around the FSW, global forces that act on the weld are calculated and are considered similar to a surface stress in front of the weld toe. In addition, structural stresses are largely mesh size insensitive in the vicinity of stress concentrations due to self-equilibrating stress distribution [56]. Near the weld toe, the structural stresses are represented by membrane and bending components that are equivalent to the local stress distribution near the weld. In many cases, the FSW joint under consideration is constructed of sheet metal and thus the use of shell elements is appropriate, which further reduces the calculation time. The reader is directed to the work of Lee et al. [57] for more details on structural stress approaches for welded joints.

To calculate the structural stresses in overlap joints, the moments and forces at elements along the centerline of the FSW can be calculated by employing elastic finite element analysis [55]. The calculations of structural stresses can be expressed as: [58]

$$\sigma_s^{(i)}(y) = \sigma_m^{(i)}(y) + \sigma_b^{(i)}(y) \tag{5.6}$$

where $\sigma_s^{(i)}(y)$ is the structural stress along the FSW, $\sigma_m^{(i)}(y)$ is the membrane stress, and $\sigma_b^{(i)}(y)$ is the bending stress. Note that the membrane and bending stresses are taken from finite element analysis similar to the model shown in Fig. 5.6. We further note that it is common in the automotive industry to represent cast components with solid elements and sheet metal with shell elements as shown in Fig. 5.6. Recall that the notch strain-life analysis performed by Wang et al. [55] presented in the previous section (5.3) resulted in nonconservative predictions of the FSLW joint in the high-cycle regime. In the same study, the structural stresses were shown to better collapse the results within a factor of 3 scatters (Fig. 5.7) as compared to the notch strain-life approach [55].

Although the structural stress approach can be used to capture different loading configurations on FSW joints, the structural stress approach can also capture macroscale defects such as keyholes that can be left in the structure if refilling techniques are not employed. As

(A)   (B)

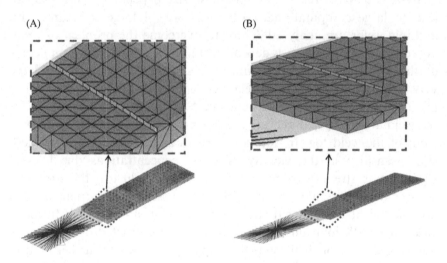

Figure 5.6 Details of finite element models for (A) FSLW lap-shear coupon and (B) FSLW lap-shear coupon with keyhole [59].

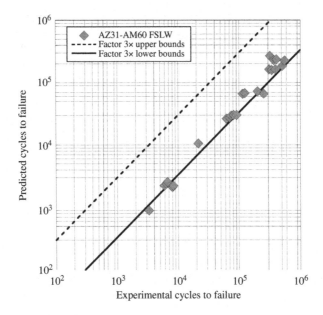

Figure 5.7 Comparison of predicted life versus experimental life by employing a structural stress approach for FSLW lap-joint coupons. Data replotted from Wang R., Kang H.-T., (Cindy) Jiang C. Fatigue life prediction for overlap friction stir linear welds of magnesium alloys. J Manuf Sci Eng 2016;138:061013. doi:10.1115/1.4032469.

an example, in a study by Rao et al. [59], structural stresses were used to capture the effect of the keyhole in FSLW joints subjected to fatigue. In their study, structural stresses were calculated for FSLW coupons with and without keyhole features. To capture the effect of the keyhole, Rao et al. modeled the keyhole region with no joint connection.

For FSLW joints, the structural stress can be calculated using the following expression [59]:

$$\sigma_\perp = -\frac{12 \cdot m_y \cdot z}{t^3} - \frac{n_x}{t} \qquad (5.7)$$

where $\sigma_\perp$ is the normal stress that acts perpendicular to the FSW line, m_y is a line moment, n_x is a line force, z is the distance from the mean surface in the local z-direction, t is the sheet thickness. The structural stress calculated using the finite element model is plotted with the fatigue life data obtained from the lab results and is presented in Fig. 5.8. Rao et al. concluded that the difference in fatigue life observed due to the presence of a keyhole feature in their study was

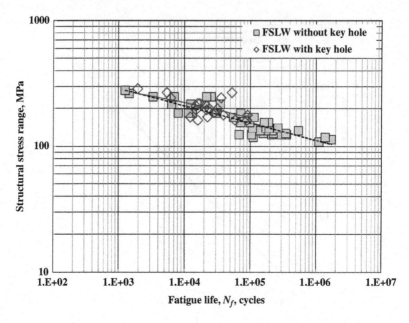

Figure 5.8 Comparison of structural stresses in friction stir linear welding coupons with and without keyhole features [59].

well captured by using structural stresses. Overall, the structural stress plot (Fig. 5.8) collapsed the FSLW joints with and without the keyhole feature into a single master curve. Thus, the results of the FEA study support the experimental observations that the keyhole reduces the effective weldment length but do not generally reduce the fatigue behavior of the joint. Furthermore, Rao et al. demonstrate the usefulness of the structural stress approach in modeling the effect the keyhole feature.

5.5 DAMAGE TOLERANCE

Damage-tolerant design is a commonly used method for engineering structures that can be inspected for cracks and have a useful life once a crack is detected. Many civil structures, aircraft fuselages, pipelines, and automotive chassis are good examples of the many types of structures that are engineered by the use of damage-tolerant design. In addition, damage tolerance can be used to assess the fatigue life of FSW by considering specific features and/or defects as the initial flaw. The following discussion regarding the use of fracture mechanics to model the fatigue behavior in FSW builds on the brief presentation of fracture mechanics concepts presented in Chapter 4, Fatigue Crack Growth in Friction Stir Welds. For additional details regarding the use LEFM to model fatigue damage, the reader is referred elsewhere [42,53].

Considering an FSW joint subjected to cyclic loading, several assumptions regarding the initial flaw size are required. As such, if one knows the initial flaw size (a_0) and can determine a critical flaw size beyond which repairs are critical (a_f), Eq. 4.11 (see Chapter 4: Fatigue Crack Growth in Friction Stir Welds) may be rearranged and integrated to determine the cycles from a known flaw size (N_0) to critical flaw size (N_f), by

$$\int_{N_0}^{N_f} dN = \int_{a_0}^{a_f} \frac{1}{b\Delta K^c} da. \tag{5.8}$$

Note that there may not be a closed-form relationship for ΔK for a particular geometry and loading condition of interest. Furthermore, closed-form solutions of ΔK for simple geometry and loading conditions, such as the CT specimen, may require piecewise integration given that ΔK is typically not linearly dependent upon the crack length, a. Finally, given the form of the Paris equation (Eq. 4.11), one may create a piecewise integration in which fatigue crack growth (FCG) in the base material, heat affected zone, thermo mechanically affected zone, and/or nugget of an FSW joint are summed when based on the engineers' understanding of critical crack path and experimental results of various FCG test arrangements as summarized in Chapter 4, Fatigue Crack Growth in Friction Stir Welds.

An example of using LEFM approach to model fatigue is presented in the work by Moraes et al. [60], where they modeled the effect of the overlap orientation in FSLW lap-shear coupons. In their study, they used a commercial code FRANC2D to calculate the stress intensity factors (SIFs) for the advancing side (AS) and retreating side (RS) features under plane strain conditions. The cracks were propagated automatically using a predefined crack extension and automatic remeshing functions contained in the FRANC2D solver. An example of the initial crack used in the crack growth calculations for the AS is shown in Fig. 5.9A, and a magnified view of the crack tip is shown in Fig. 5.9B. An example of the final crack extension with the remeshing around the crack as it grew through the AS-orientated sample is shown in

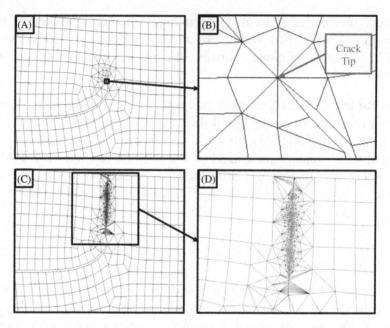

Figure 5.9 (A) Mesh of the initial crack of advancing side coupon and (B) a magnified view of the crack tip. (C) Overall view of the mesh of the final crack, and (D) a magnified view of the propagated crack [60].

Fig. 5.9C, where a magnified view of the final crack shown in Fig. 5.9D. For the RS, the meshing of the initial crack and the crack extension was similar to the AS but is not shown for brevity's sake. Fig. 5.10A shows the deformed mesh at the final crack extension, and Fig. 5.10B shows the corresponding Mode I stress intensity (KI) versus crack length for the AS of the weld. Similar results were generated for the RS of the weld. We note that in the work by Moraes et al. [60], they used FRANC2D software, but other commercial softwares like ABAQUS and ANYSIS can provide similar analysis.

To calculate the fatigue life, Moraes et al. [60] used the Paris relationship to estimate the number of cycles to failure for the AS and RS stacking orientation. In their work, they assumed the number of cycles to initiate the fatigue crack was negligible, and thus their use of a crack growth-only approach seems reasonable. Because the SIF was calculated as a function of crack length for each crack increment using FRANC2D, which in this study was $\Delta a = 2.54 \times 10^{-2}$ mm, the calculation of the number of cycles was achieved by integrating Eq. 4.11 (see Chapter 4: Fatigue Crack Growth in Friction Stir Welds).

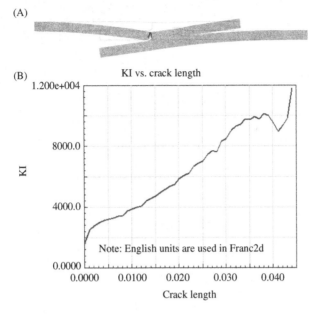

Figure 5.10 An example of the (A) deformed mesh and (B) the corresponding calculation of Mode I stress intensity (KI) as a function of crack length for an FSLW lap-shear coupon.

$$N_i = \sum_0^i \frac{(a_i - a_{i-1})}{C[(K_i - K_{i-1})/2]^m} \qquad (5.9)$$

where i is the crack increment, and K is the SIF, a is the crack length, N is the number of cycles, and C and m are material constants.

Fig. 5.11 presents the comparison of LEFM approach to the experimental results of the AS- and RS-orientated FSLW coupons. Although not an exact correlation between the model and the experimental results was achieved, the results by Moraes et al. [60] make a strong case for using a damage tolerance approach, because the observations of the cracking behavior in FSW aligns well with mechanics of LEFM. However, Moraes et al. [60] acknowledge that their modeling efforts could be improved by including mixed mode (I + II) as opposed to just Mode I, and by also including parameters, C and m, which represent FCG in the region of the weld zone. However, obtaining C and m properties for cracks that propagate in the particular location of a weld as in an FSLW lap-shear coupon is likely difficult. In any case, the results of using LEFM for modeling the fatigue life of this particular lap-joint suggests that the difference in observed performance between

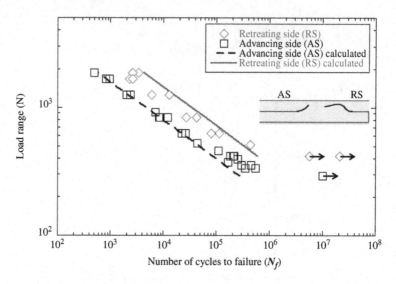

Figure 5.11 Comparison of the experimental fatigue results to the calculated number of cycle to failure of the AS-and RS-oriented FSLW coupons [60].

the AS and RS coupons is a first-order effect of the geometrical features of the FSLW.

In the study by Moraes et al. [60], the stress intensity solution was calculated as a function of crack length. However, in some cases, the use of closed-form solutions have also shown good results in modeling the fatigue behavior in lap-joints. For example, many studies [61−69] have examined the use of a closed-form SIF solution in modeling the fatigue behavior FSSW lap-joints. Of course, similar to the work by Moraes et al., joint specific stress intensity solutions have been calculated for FSW overlap joints [70−73].

In modeling the FSSW with an LEFM approach, one can consider the FCG in the coupon as a planar kinked crack. First proposed by Newman and Dowling [74], the kinked crack method is based on the assumption that the fatigue crack in the FSSW behaves similar to a kinked crack, and thus the global Mode I and II SIF can be correlated to a local SIF as a function of kinked crack angle θ [75,76]:

$$k_I = \frac{K_I}{4}\left[3cos\left(\frac{\theta}{2}\right) + cos\left(\frac{3\theta}{2}\right)\right] - \frac{K_{II}}{4}\left[3sin\left(\frac{\theta}{2}\right) + 3sin\left(\frac{3\theta}{2}\right)\right] \quad (5.10)$$

$$k_{II} = \frac{K_I}{4}\left[sin\left(\frac{\theta}{2}\right) + sin\left(\frac{3\theta}{2}\right)\right] + \frac{K_{II}}{4}\left[cos\left(\frac{\theta}{2}\right) + 3cos\left(\frac{3\theta}{2}\right)\right] \quad (5.11)$$

where k_I and k_{II} are the local SIFs, K_I and K_{II} are the global SIFs, and θ is the angle of the kinked crack. To calculate the number cycles to failure, an equivalent mode I SIF is used [77]:

$$k_{eq} = \sqrt{k_I^2 + k_{II}^2} \quad (5.12)$$

There are several alternatives of the equivalent Mode I SIF expression, including some that contain a correction factor that capture the sensitivity of materials to Mode II loading [78]. However, it is not clear what value this correction factor should be for overlap lap-shear loading conditions [69]. Integrating Eq. 4.11 (see Chapter 4: Fatigue Crack Growth in Friction Stir Welds), the number of cycles to failure, N_{Total}, for an FSSW can be calculated:

$$N_{Total} = \frac{t - t_{crack}}{Csin\theta}\left(\Delta k_{eq}\right)^{-m} \quad (5.13)$$

Additional modifications to the LEFM approach to model specific low-cycle and high-cycle regimes of the fatigue behavior of the FSSW overlap joints are presented, for example, in work by Lin et al. [67,79]. An example on the use of the LEFM approach on AA2024 FSSW lap-shear coupons is shown in Fig. 5.12, where in this study, the authors considered several LEFM approaches (denoted as separate equations in the legend of Fig. 5.12). Please note that the equations shown in Fig. 5.12 refer to the equations presented in [79].

Before concluding the presentation of the LEFM approach in modeling fatigue in FSW, it is prudent to briefly discuss the use of LEFM in helping to understand the effect of welding parameters on fatigue properties. Because the mechanics of the LEFM approach is a function of the geometry of the FSW joint, the loading configuration, and FCG parameters, it is conceivable that LEFM could be used to determine higher order effects of welding parameters on fatigue life. An example of the use of LEFM to assess the effect of the structural integrity of FSSW is presented in the work by Jordon [80]. In this work, the effect of sheet thickness, interfacial hook height, and nugget diameter on simulated fatigue behavior in FSSW joints of AZ31 Mg alloy were examined, as shown in Fig. 5.13. Although the effective

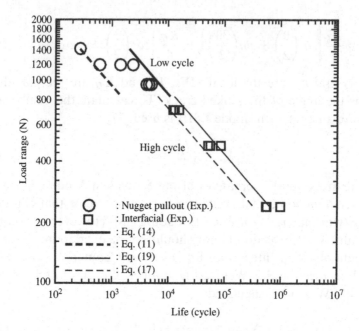

Figure 5.12 Example of the use of kinkedcrack LEFM approach to predict fatigue in FSLW lap joints [79].

sheet thickness had the greatest effect on the fatigue life, the size of the interfacial hook, which is strongly influenced by the welding specific parameters discussed in Chapter 3, Influence of Welding Parameters on Fatigue Behavior, was also found to strongly influence the number of cycles to failure. Although it is already well known to the FSW community that the hooking size is important in determining the strength of FSSW coupon, this modeling example illustrates the potential of LEFM to optimize joints for superior fatigue resistance.

5.6 MICROSTRUCTURE-SENSITIVE MODELING

We conclude this chapter on modeling of fatigue in FSW by a discussion of the microstructure-sensitive multistage fatigue (MSF) approach. Initially proposed by McDowell et al. [81] to model the fatigue performance of a cast aluminum alloy A356, the MSF model has since been adapted and expanded to predict fatigue behavior for a wide range of materials, material processing conditions [82—87], and more recently to FSBW [43,44,47,88] and FSSW [89] joints. Although the previous sections in this chapter provided an overview of common modeling

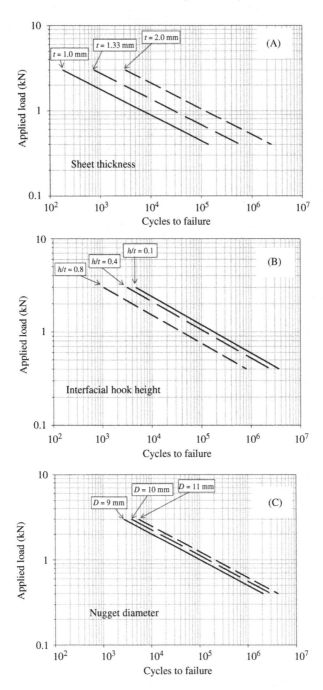

Figure 5.13 Effect of the geometrical features on fatigue life of FSSW joints: (A) sheet thickness, (B) interfacial hook height, and (C) nugget diameter. Modeling results show that the LEFM approach can predict the sensitivities related to FSSW. Adapted from Jordon JB. The effect of microstructural and geometrical features on fatigue performance in MG AZ31 friction stir spot welds. ASME 2011 International Mechanical Engineering Congress and Exposition, IMECE 2011, vol. 8; 2011.

approaches employed for FSW, the presentation of the MSF model is provided in detail for better understanding.

The MSF model was originally developed in conjunction with Sandia National Laboratories [81] to evaluate the sensitivity of the fatigue response to microstructural features with the purpose of fatigue life prediction in the design of materials and structural components. As such, the model considers the role of local constrained microplasticity at inclusions and their effect in the crack incubation, and the microstructural small crack growth and physically small crack growth [86]. Lastly, based on the variation of microstructure and material defects, the upper and lower bounds of the fatigue life can be predicted. The model has since then been adapted to consider crack incubation from microscale discontinuities in FSW such as voids [44], intermetallic particles [43,47,88], and oxide debris [89].

The fatigue damage evolution predicted by the MSF model is divided in three main stages as:

$$N_{Total} = N_{inc} + N_{MSC/PSC} + N_{LC} \tag{5.14}$$

where N_{Total} is the total fatigue life, N_{inc} is the number of cycles required for crack incubation, $N_{MSC/PSC}$ and N_{LC} are the number of cycles required for the propagation of the microstructurally small/physically small crack (MSC/PSC), and the propagation of the long crack (LC), respectively.

For FSW applications, the N_{inc} comprises the number of cycles of a crack incubating at some discontinuity such as intermetallic particles or macroscale defects like volumetric voids. The incubation stage in the MSF model is correlated to a damage parameter in a modified Coffin–Manson law. This nonlocal parameter is expressed as:

$$C_{inc}N_{inc}^{\alpha} = \beta \tag{5.15}$$

$$C_{inc} = CNC + z(C_m - CNC) \tag{5.16}$$

$$CNC = C_n(1 - R) \tag{5.17}$$

where β is the nonlocal damage term, C_{inc} and α are the coefficient and exponent values obtained for the modified Coffin–Manson law for incubation, respectively (Eq. 5.15). On the other hand, C_m and C_n

are model constants and the R and z are the load ratio and localization multiplier, respectively (Eq. 5.18).

$$z = \frac{\frac{l}{D} - \eta_{lim}}{1 - \eta_{lim}} \qquad (5.18)$$

Thus, D is the size of the critical inclusion where that crack incubates, l is the size of the plastic zone in front of the discontinuity, and η_{lim} is the limiting factor that defines the transition from constrained to unconstrained micro-notch root plasticity. Note that β is estimated by the following relations:

$$\beta = \frac{\Delta\gamma_{max}^{p*}}{2} = Y[\varepsilon_a - \varepsilon_{th}]^q, \frac{l}{D} < \eta_{lim} \qquad (5.19)$$

$$\beta = \frac{\Delta\gamma_{max}^{p*}}{2} = Y(1 + \zeta z)[\varepsilon_a - \varepsilon_{th}]^q, \frac{l}{D} > \eta_{lim} \qquad (5.20)$$

Here, ε_a is the remote applied strain amplitude, ε_{th} is the microplasticity threshold, and l is the nominal linear dimension of the plastic zone size in front of the inclusion. The ratio l/D is defined as the square root of the ratio of the plastic zone over the inclusion area, and the parameters q and ζ are determined from micromechanical simulations [86,90]. The limiting ratio, η_{lim}, indicates the transition from proportional (constrained) micro-notch root plasticity to nonlinear (unconstrained) micro-notch root plasticity with respect to the applied strain amplitude, where $\eta_{lim} = 0.3$ has been found to be suitable for FSW joints [47,88]. The parameter Y [47,88] is expressed as $Y = y_1 + (1 + R)y_2$, where R is the load ratio, and y_1 and y_2 are model constants. For completely reversed loading cases, $Y = y_1$. Furthermore, when l/D reaches its limits, the parameter Y is revised to include the geometric effects related to the type of inclusion, $\overline{Y} = [1 + (l/D)]Y$. The correlation of the plastic zone size to number of cycles is calculated using the expression of $(\Delta\gamma_{max}^{p*})/2$ with respect to the remote loading strain amplitude,

$$\frac{l}{D} = \eta_{lim} \frac{\langle\varepsilon_a - \varepsilon_{th}\rangle}{\varepsilon_{per} - \varepsilon_{th}}, \frac{l}{D} \leq \eta_{lim}, \qquad (5.21)$$

$$\frac{l}{D} = 1 - (1 - \eta_{lim})\left(\frac{\varepsilon_{per}}{\varepsilon_a}\right)^r, \frac{l}{D} > \eta_{lim}, \qquad (5.22)$$

where r, a shape constant for the transition to the limited plasticity is determined through micromechanical simulations [86,90], and ε_{er} is the percolation limit.

The driving force for the propagation of microstructurally/physically small cracks in $N_{MSC/PSC}$ stage for FSW is defined by the crack tip opening displacement and is expressed as

$$\left(\frac{da}{dN}\right)_{MSC} = \chi(\Delta CTD - \Delta CTD_{th}), \tag{5.23}$$

where ΔCTD is the crack tip opening displacement range, ΔCTD_{th} is the crack tip displacement threshold range, and χ is a material constant. The crack tip opening displacement range is defined as

$$\Delta CTD = C_{II}\left(\frac{GS}{GS_0}\right)^{\omega}\left(\frac{GO}{GO_0}\right)^{\varpi}\left[\frac{U\Delta\hat{\sigma}}{S_{ut}}\right]^{\zeta} a$$
$$+ C_I\left(\frac{GS}{GS_0}\right)^{\omega}\left(\frac{GO}{GO_0}\right)^{\varpi}\left(\frac{\Delta\gamma^P_{max}}{2}\right)^2 \tag{5.24}$$

where C_I is the low-cycle fatigue coefficient, C_{II} and ζ are the coefficient and exponents for the high-cycle fatigue regime, the initial crack length a, and S_{ut} is the ultimate tensile strength obtained from the monotonic tensile test. The GS, GS_0 GO, GO_0, ω, and ϖ parameters are model constants for grain size and grain orientation, which may be deactivated by setting the corresponding exponent to zero. The equivalent uniaxial stress amplitude, $\Delta\hat{\sigma} = \overline{\sigma}_a + (1 - \theta)\Delta\sigma_1$, is defined as the linear combination of effective stress amplitude, $\overline{\sigma}_a = \left[(3/2)(\Delta\sigma'_{ij}/2)(\Delta\sigma'_{ij}/2)\right]^{0.5}$, and the maximum principal stress range $\Delta\sigma_1$, where $0 \leq \theta \leq 1$ as the loading parameter. For mean stress effects on crack growth, the U parameter is employed, where $U = [1/(1 - R)]$ is for the case when $R \leq 0$ and $U = 1$ for $R > 0$. Once the driving force of the fatigue crack is high enough within the FSW joint, FCG rate is defined by LEFM, where established expressions for LC growth (Eq. 5.8) can be used to model the remaining life similar to work by Xue et al. [91].

An example of the adaption of the MSF model to FSW is presented in Fig. 5.14. In this work, Rodriguez et al. [47] adapted the MSF model to predict the incubation and small fatigue crack stages of fatigue in the dissimilar FSBW of AA6061-to-AA7050, as shown in

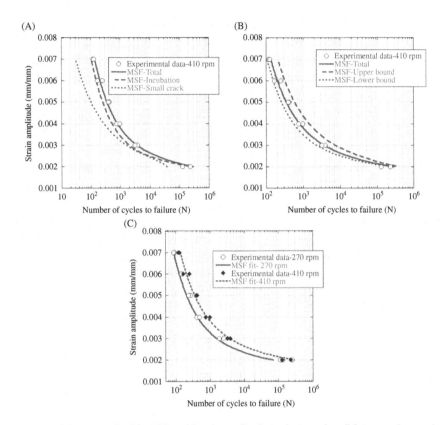

Figure 5.14 (A) An example of the MSF model used to predict the incubation and small fatigue crack stages for dissimilar FSW of AA6061-to-AA7050 and (B) the MSF correlation to the experimental results including the upper bound and lower bounds. (C) Comparison of the MSF correlation obtained between two rotational speeds (270 and 410 rpm) [47].

Fig. 5.14A. In addition, Fig. 5.14B shows the comparison of the MSF model to the upper and lower fatigue life bounds, which were based ranges of inclusion sizes determined from metallographic analysis. In Fig. 5.14C, the MSF prediction is compared to the fatigue behavior of the two different welding parameters.

Further evolution of the MSF model has included modifying the incubation expression to capture the relationship between hardness and grain size in FSBW of AA2099:

$$\varpi = \left(\frac{HV/GS}{HV_o/GS_0} \right)^{j} \tag{5.25}$$

where HV and GS are the hardness and grain size, respectively. In addition, the parameters HV_o and GS_0 are reference states for the hardness and grain sizes, respectively, and the parameter j represents a material dependent parameter. Fig. 5.15 shows the use of the MSF model with the adaption of the dimensionless parameter, where the model shows good correlation to the fatigue life of the base material and FSW joints made from two different sets of welding parameters [43]. The motivation of including this new parameter is to address competing mechanisms, such as the contrast between hard and soft regions of the FSW zones, and variation in grain sizes that do not always follow a Hall–Petch relationship. Although future work is still needed to fully validate the inclusion of Eq. 5.25, the presentation of the dimensionless parameter further illustrates the use of the MSF model in predicting the variation in fatigue life of the FSW.

Lastly, the MSF model was adapted by Rao and Jordon [92] to capture fatigue in FSSW joints. The adaption of the MSF model to FSSW joints required modifying several aspects of the model. Although the incubation, small crack, and LC framework remained unchanged, specific equations within the model were modified to account for the changes in the mechanics of fatigue in lap joints. In particular, the LC stage was modified to capture the experimentally observed kinked crack behavior. The modifications to the LC stage

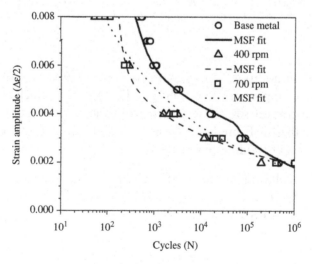

Figure 5.15 Comparison of MSF model with the addition of the hardness and grain size nondimensionless parameter to the experimental results for AA2099 [43].

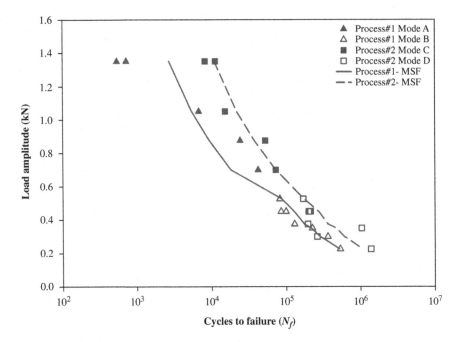

Figure 5.16 Example of the MSF model applied to FSSW lap-joint for two sets of welding parameters [89]. Reprinted with permission from Springer Nature.

included similar expressions as presented earlier in the modeling of the FSSW using LEFM approaches (Section 5.5), whereas the incubation and small crack stages of the MSF model remained unchanged and thus retained the microstructure sensitivities of the original model. Fig. 5.16 shows the MSF model correlated to the fatigue behavior of FSSW of AZ31 Mg alloy for two sets of welding parameters.

REFERENCES

[1] Standard B. Eurocode 9—Design of aluminium structures; 2007.

[2] Hobbacher A. Recommendations for fatigue design of welded joints and components. IIW Doc. XIII-2151r1-07/XV-1254r1-07 2007.

[3] Maddox SJ. Review of fatigue assessment procedures for welded aluminium structures. Int J Fatigue 2003;25:1359−78. Available from: https://doi.org/10.1016/S0142-1123(03)00063-X.

[4] Yan Z, Liu X, Fang H. Mechanical properties of friction stir welding and metal inert gas welding of Al-Zn aluminum alloy joints. Int J Adv Manuf Technol 2017;91:3025−31. Available from: https://doi.org/10.1007/s00170-017-0021-y.

[5] Baragetti S, D'Urso G. Aluminum 6060-T6 friction stir welded butt joints: Fatigue resistance with different tools and feed rates. J Mech Sci Technol 2014;28:867−77. Available from: https://doi.org/10.1007/s12206-013-1152-1.

[6] Deng C, Gao R, Gong B, Yin T, Liu Y. Correlation between micro-mechanical property and very high cycle fatigue (VHCF) crack initiation in friction stir welds of 7050 aluminum alloy. Int J Fatigue 2017;104:283−92. Available from: https://doi.org/10.1016/j.ijfatigue.2017.07.028.

[7] Dubourg L, Merati A, Jahazi M. Process optimisation and mechanical properties of friction stir lap welds of 7075-T6 stringers on 2024-T3 skin. Mater Des 2010;31:3324−30. Available from: https://doi.org/10.1016/j.matdes.2010.02.002.

[8] Gungor B, Kaluc E, Taban E, Sik A. Mechanical, fatigue and microstructural properties of friction stir welded 5083-H111 and 6082-T651 aluminum alloys. Mater Des 2014;56:84−90. Available from: https://doi.org/10.1016/j.matdes.2013.10.090.

[9] He C, Yang K, Liu Y, Wang Q, Cai M. Improvement of very high cycle fatigue properties in an AA7075 friction stir welded joint by ultrasonic peening treatment. Fatigue Fract Eng Mater Struct 2017;40:460−8. Available from: https://doi.org/10.1111/ffe.12516.

[10] He C, Kitamura K, Yang K, Liu YJ, Wang QY, Chen Q. Very high cycle fatigue crack initiation mechanism in nugget zone of AA 7075 friction stir welded joint. Adv Mater Sci Eng 2017;3:1−10. Available from: https://doi.org/10.1155/2017/7189369.

[11] Kadlec M, Ruzek R, Novakova L. Mechanical behaviour of AA 7475 friction stir welds with the kissing bond defect. Int J Fatigue 2015;74:7−19. Available from: https://doi.org/10.1016/j.ijfatigue.2014.12.011.

[12] Mahdavi Shahri M, Sandström R. Effective notch stress and critical distance method to estimate the fatigue life of T and overlap friction stir welded joints. Eng Fail Anal 2012;25:250−60. Available from: https://doi.org/10.1016/j.engfailanal.2012.05.019.

[13] Minak G, Ceschini L, Boromei I, Ponte M. Fatigue properties of friction stir welded particulate reinforced aluminium matrix composites. Int J Fatigue 2010;32:218−26. Available from: https://doi.org/10.1016/j.ijfatigue.2009.02.018.

[14] Moreira PMGP, de Jesus AMP, de Figueiredo MAV, Windisch M, Sinnema G, de Castro PMST. Fatigue and fracture behaviour of friction stir welded aluminium-lithium 2195. Theor Appl Fract Mech 2012;60:1−9. Available from: https://doi.org/10.1016/j.tafmec.2012.06.001.

[15] Moreira PMGP, de Oliveira FMF, de Castro PMST. Fatigue behaviour of notched specimens of friction stir welded aluminium alloy 6063-T6. J Mater Process Technol 2008;207:283−92. Available from: https://doi.org/10.1016/j.jmatprotec.2007.12.113.

[16] Barsoum Z, Khurshid M, Barsoum I. Fatigue strength evaluation of friction stir welded aluminium joints using the nominal and notch stress concepts. Mater Des 2012;41:231−8. Available from: https://doi.org/10.1016/j.matdes.2012.05.018.

[17] Nelaturu P, Jana S, Mishra RS, Grant G, Carlson BE. Influence of friction stir processing on the room temperature fatigue cracking mechanisms of A356 aluminum alloy. Mater Sci Eng A 2018;716:165−78. Available from: https://doi.org/10.1016/j.msea.2018.01.044.

[18] Ni DR, Chen DL, Xiao BL, Wang D, Ma ZY. Residual stresses and high cycle fatigue properties of friction stir welded SiCp/AA2009 composites. Int J Fatigue 2013;55:64−73. Available from: https://doi.org/10.1016/j.ijfatigue.2013.05.010.

[19] Sharma C, Dwivedi DK, Kumar P. Fatigue behavior of friction stir weld joints of Al-Zn-Mg alloy AA7039 developed using base metal in different temper condition. Mater Des 2014;64:334−44. Available from: https://doi.org/10.1016/j.matdes.2014.07.013.

[20] Sillapasa K, Surapunt S, Miyashita Y, Mutoh Y, Seo N. Tensile and fatigue behavior of SZ, HAZ and BM in friction stir welded joint of rolled 6N01 aluminum alloy plate. Int J Fatigue 2014;63:162−70. Available from: https://doi.org/10.1016/j.ijfatigue.2014.01.021.

[21] Besel M, Besel Y, Alfaro Mercado U, Kakiuchi T, Uematsu Y. Fatigue behavior of friction stir welded Al-Mg-Sc alloy. Int J Fatigue 2015;77:1−11. Available from: https://doi.org/10.1016/j.ijfatigue.2015.02.013.

[22] Boni L, Lanciotti A, Polese C. "Size effect" in the fatigue behavior of friction stir welded plates. Int J Fatigue 2015;80:238−45. Available from: https://doi.org/10.1016/j.ijfatigue.2015.06.010.

[23] Cavaliere P, Cabibbo M, Panella F, Squillace A. 2198 Al-Li plates joined by friction stir welding: mechanical and microstructural behavior. Mater Des 2009;30:3622−31. Available from: https://doi.org/10.1016/j.matdes.2009.02.021.

[24] Cavaliere P, De Santis A, Panella F, Squillace A. Effect of anisotropy on fatigue properties of 2198 Al-Li plates joined by friction stir welding. Eng Fail Anal 2009;16:1856−65. Available from: https://doi.org/10.1016/j.engfailanal.2008.09.024.

[25] Cavaliere P, De Santis A, Panella F, Squillace A. Effect of welding parameters on mechanical and microstructural properties of dissimilar AA6082-AA2024 joints produced by friction stir welding. Mater Des 2009;30:609−16. Available from: https://doi.org/10.1016/j.matdes.2008.05.044.

[26] da Silva J, Costa JM, Loureiro A, Ferreira JM. Fatigue behaviour of AA6082-T6 MIG welded butt joints improved by friction stir processing. Mater Des 2013;51:315−22. Available from: https://doi.org/10.1016/j.matdes.2013.04.026.

[27] Deng C, Wang H, Gong B, Li X, Lei Z. Effects of microstructural heterogeneity on very high cycle fatigue properties of 7050-T7451 aluminum alloy friction stir butt welds. Int J Fatigue 2016;83:100−8. Available from: https://doi.org/10.1016/j.ijfatigue.2015.10.001.

[28] Padmanaban G, Balasubramanian V. An experimental investigation on friction stir welding of AZ31B magnesium alloy. Int J Adv Manuf Technol 2010;49:111−21. Available from: https://doi.org/10.1007/s00170-009-2368-1.

[29] Padmanaban G, Balasubramanian V. Fatigue performance of pulsed current gas tungsten arc, friction stir and laser beam welded AZ31B magnesium alloy joints. Mater Des 2010;31:3724−32. Available from: https://doi.org/10.1016/j.matdes.2010.03.013.

[30] Zhou L, Li ZY, Nakata K, Feng JC, Huang YX, Liao JS. Microstructure and fatigue behavior of friction stir-welded noncombustive Mg-9Al-Zn-Ca magnesium alloy. J Mater Eng Perform 2016;25:2403−11. Available from: https://doi.org/10.1007/s11665-016-2061-0.

[31] Logan BP, Toumpis AI, Galloway AM, McPherson NA, Hambling SJ. Dissimilar friction stir welding of duplex stainless steel to low alloy structural steel. Sci Technol Weld Join 2016;21:11−19. Available from: https://doi.org/10.1179/1362171815Y.0000000063.

[32] Polezhayeva H, Toumpis AI, Galloway AM, Molter L, Ahmad B, Fitzpatrick ME. Fatigue performance of friction stir welded marine grade steel. Int J Fatigue 2015;81:162−70. Available from: https://doi.org/10.1016/j.ijfatigue.2015.08.003.

[33] Edwards P, Ramulu M. Identification of process parameters for friction stir welding Ti−6Al−4V. J Eng Mater Technol 2010;132:031006. Available from: https://doi.org/10.1115/1.4001302.

[34] Jana S, Hovanski Y. Fatigue behaviour of magnesium to steel dissimilar friction stir lap joints. Sci Technol Weld Join 2012;17:141−5. Available from: https://doi.org/10.1179/1362171811Y.0000000083.

[35] Miranda AC, de O, Gerlich A, Walbridge S. Aluminum friction stir welds: review of fatigue parameter data and probabilistic fracture mechanics analysis. Eng Fract Mech 2015;147:243−60. Available from: https://doi.org/10.1016/j.engfracmech.2015.09.007.

[36] Dickerson TL, Przydatek J. Fatigue of friction stir welds in aluminium alloys that contain root flaws. Int J Fatigue 2003;25:1399−409. Available from: https://doi.org/10.1016/S0142-1123(03)00060-4.

[37] Vigh LG, Okura I. Fatigue behaviour of friction stir welded aluminium bridge deck segment. Mater Des 2013;44:119−27. Available from: https://doi.org/10.1016/j.matdes.2012.08.007.

[38] Zhou C, Yang X, Luan G. Effect of root flaws on the fatigue property of friction stir welds in 2024-T3 aluminum alloys. Mater Sci Eng A 2006;418:155–60. Available from: https://doi.org/10.1016/j.msea.2005.11.042.

[39] Di SS, Yang XQ, Luan GH, Jian B. Comparative study on fatigue properties between AA2024-T4 friction stir welds and base materials. Mater Sci Eng a-Structural Mater Prop Microstruct Process 2006;.

[40] Shahri MM, Sandström R. Fatigue analysis of friction stir welded aluminium profile using critical distance. Int J Fatigue 2010;32:302–9. Available from: https://doi.org/10.1016/j.ijfatigue.2009.06.019.

[41] Mahdavi Shahri M, Sandström R, Osikowicz W. Critical distance method to estimate the fatigue life time of friction stir welded profiles. Int J Fatigue 2012;37:60–8. Available from: https://doi.org/10.1016/j.ijfatigue.2011.10.003.

[42] Bannantine JA, Comer JJ, Handrock JL. Fundamentals of metal fatigue analysis, vol. 90. Englewood Cliffs, NJ: Prentice Hall; 1990.

[43] Cisko AR. Fatigue characterization and microstructure-sensitive modeling of extruded and friction stir welded aluminum lithium alloy 2099. Tuscaloosa, AL: The University of Alabama; 2018.

[44] Cisko A.R., Jordon B., Rodriguez R., Rao H., Allison P.G. Microstructure-sensitive fatigue modeling of friction stir welded aluminum alloy 6061. The ASME 2015 International Mechanical Engineering Congress and Exposition, American Society of Mechanical Engineers; 2015, V014T11A003-V014T11A003.

[45] Xu WF, Liu JH, Chen DL, Luan GH. Low-cycle fatigue of a friction stir welded 2219-T62 aluminum alloy at different welding parameters and cooling conditions. Int J Adv Manuf Technol 2014;74:209–18. Available from: https://doi.org/10.1007/s00170-014-5988-z.

[46] Ceschini L, Boromei I, Minak G, Morri A, Tarterini F. Effect of friction stir welding on microstructure, tensile and fatigue properties of the AA7005/10 vol.%Al2O3p composite. Compos Sci Technol 2007;67:605–15. Available from: https://doi.org/10.1016/j.compscitech.2006.07.029.

[47] Rodriguez RII, Jordon JBB, Allison PGG, Rushing T, Garcia L. Low-cycle fatigue of dissimilar friction stir welded aluminum alloys. Mater Sci Eng A 2016;654:236–48. Available from: https://doi.org/10.1016/j.msea.2015.11.075.

[48] Feng AH, Chen DL, Ma ZY. Microstructure and low-cycle fatigue of a friction-stir-welded 6061 aluminum alloy. Metall Mater Trans A Phys Metall Mater Sci 2010;41:2626–41. Available from: https://doi.org/10.1007/s11661-010-0279-2.

[49] White BC, Rodriguez RI, Cisko A, Jordon JB, Allison PG, Rushing T, et al. Effect of heat exposure on the fatigue properties of AA7050 friction stir welds. J Mater Eng Perform 2018;27. Available from: https://doi.org/10.1007/s11665-018-3379-6.

[50] Wang SQ, Liu JH, Chen DL. Strain-controlled fatigue properties of dissimilar welded joints between Ti-6Al-4V and Ti17 alloys. Mater Des 2013;49:716–27. Available from: https://doi.org/10.1016/j.matdes.2013.02.034.

[51] Ni DR, Chen DL, Yang J, Ma ZY. Low cycle fatigue properties of friction stir welded joints of a semi-solid processed AZ91D magnesium alloy. Mater Des 2014;56:1–8. Available from: https://doi.org/10.1016/j.matdes.2013.10.081.

[52] Yang J, Ni DR, Wang D, Xiao BL, Ma ZY. Strain-controlled low-cycle fatigue behavior of friction stir-welded AZ31 magnesium alloy. Metall Mater Trans A Phys Metall Mater Sci 2014;45:2101–15. Available from: https://doi.org/10.1007/s11661-013-2129-5.

[53] Stephens RI, Fatemi A, Stephens RR, Fuchs HO. Metal Fatigue in Engineering. New York: John Wiley & Sons; 2000.

[54] Sun G, Chen Y, Chen S, Shang D. Fatigue modeling and life prediction for friction stir welded joint based on microstructure and mechanical characterization. Int J Fatigue 2017;98:131−41. Available from: https://doi.org/10.1016/j.ijfatigue.2017.01.025.

[55] Wang R, Kang H-T, (Cindy) Jiang C. Fatigue life prediction for overlap friction stir linear welds of magnesium alloys. J Manuf Sci Eng 2016;138:061013. Available from: https://doi.org/10.1115/1.4032469.

[56] Dong P. A structural stress definition and numerical implementation for fatigue analysis of welded joints. Int J Fatigue 2001;. Available from: https://doi.org/10.1016/S0142-1123(01)00055-X.

[57] Lee Y-L, Pan J, Hathaway R, Barkey M. Fatigue testing and analysis: theory and practice, vol. 13. Butterworth-Heinemann; 2005.

[58] Fermér M, Andréasson M, Frodin B. Fatigue life prediction of MAG-welded thin-sheet structures. SAE Technical Paper; 1998.

[59] Rao HM, Jordon JB, Boorgu SK, Kang H, Yuan W, Su X. Influence of the key-hole on fatigue life in friction stir linear welded aluminum to magnesium. Int J Fatigue 2017;105. Available from: https://doi.org/10.1016/j.ijfatigue.2017.08.012.

[60] Moraes JFCFC, Rodriguez RII, Jordon JBB, Su X. Effect of overlap orientation on fatigue behavior in friction stir linear welds of magnesium alloy sheets. Int J Fatigue 2017;100:1−11. Available from: https://doi.org/10.1016/j.ijfatigue.2017.02.018.

[61] Lin P-C, Chen W. Fatigue Analysis of swept friction stir clinch joints between aluminum and steel sheets. SAE Int J Mater Manuf 2017;10. Available from: https://doi.org/10.4271/2017-01-0478.

[62] Jordon JB, Horstemeyer MF, Grantham J, Badarinarayan H, State M, Products A, et al. Fatigue evaluation of friction stir spot welds in magnesium sheets. Magnes Technol 2010;2010:267−71.

[63] Su ZM, He RY, Lin PC, Dong K. Fatigue analyses for swept friction stir spot welds in lap-shear specimens of alclad 2024-T3 aluminum sheets. Int J Fatigue 2014;61:129−40. Available from: https://doi.org/10.1016/j.ijfatigue.2013.11.021.

[64] Wang DA, Chen CH. Fatigue lives of friction stir spot welds in aluminum 6061-T6 sheets. J Mater Process Technol 2009;209:367−75. Available from: https://doi.org/10.1016/j.jmatprotec.2008.02.008.

[65] Tran VX, Pan J, Pan T. Fatigue behavior of spot friction welds in lap-shear and cross-tension specimens of dissimilar aluminum sheets. Int J Fatigue 2010;32:1022−41. Available from: https://doi.org/10.1016/j.ijfatigue.2009.11.009.

[66] Tran VX, Pan J, Pan T. Fatigue behavior of aluminum 5754-O and 6111-T4 spot friction welds in lap-shear specimens. Int J Fatigue 2008;30:2175−90. Available from: https://doi.org/10.1016/j.ijfatigue.2008.05.025.

[67] Lin PC, Lo SM, Wu SP. Fatigue life estimations of alclad AA2024-T3 friction stir clinch joints. Int J Fatigue 2018;107:13−26. Available from: https://doi.org/10.1016/j.ijfatigue.2017.10.011.

[68] Lin P, Pan J, Pan T. Failure modes and fatigue life estimations of spot friction welds in lap-shear specimens of aluminum 6111-T4 sheets. Part 2: welds made by a flat tool. Int J Fatigue 2008;30:90−105. Available from: https://doi.org/10.1016/j.ijfatigue.2007.02.017.

[69] Lin P, Pan J, Pan T. Failure modes and fatigue life estimations of spot friction welds in lap-shear specimens of aluminum 6111-T4 sheets. Part 1: welds made by a concave tool. Int J Fatigue 2008;30:74−89. Available from: https://doi.org/10.1016/j.ijfatigue.2007.02.016.

[70] Tran V-X, Pan J. Analytical stress intensity factor solutions for resistance and friction stir spot welds in lap-shear specimens of different materials and thicknesses. Eng Fract Mech 2010;77:2611−39. Available from: https://doi.org/10.1016/j.engfracmech.2010.06.022.

[71] Lai W-J, Pan J. Stress intensity factor solutions for dissimilar welds in lap-shear specimens of steel, magnesium, aluminum and copper sheets. SAE Int J Mater Manuf 2015;8:563−77.

[72] Sripichai K, Pan J. Closed-form structural stress and stress intensity factor solutions for spot welds in square plates under opening loading conditions. Eng Fract Mech 2012;93:168−88. Available from: https://doi.org/10.1016/j.engfracmech.2012.06.016.

[73] Tran V-X, Pan J. Effects of weld geometry and sheet thickness on stress intensity factor solutions for spot and spot friction welds in lap-shear specimens of similar and dissimilar materials. Eng Fract Mech 2010;77:1417−38. Available from: https://doi.org/10.1016/j.engfracmech.2010.03.016.

[74] Newman J, Dowling NE. A crack growth approach to life prediction of spot-welded lap joints. Fatigue Fract Eng Mater Struct 1998;21:1123−32.

[75] Bilby BA, Cardew GE, Howard IC. Stress intensity factors at the tips of kinked and forked cracks. Fourth Int Conf Fract 1977;3A:197−200.

[76] Cotterell B, Rice JR. Slightly curved or kinked cracks. Int J Fract 1980;16:155−69.

[77] Biner SB. Fatigue crack growth studies under mixed-mode loading. Int J Fatigue 2001;23:259−63. Available from: https://doi.org/10.1016/S0142-1123(01)00146-3.

[78] Broek D. Elementary engineering fracture mechanics. Springer Netherlands; 1986.

[79] Su ZM, He RY, Lin PC, Dong K. Fatigue of alclad AA2024-T3 swept friction stir spot welds in cross-tension specimens. J Mater Process Technol 2016;236:162−75. Available from: https://doi.org/10.1016/j.jmatprotec.2016.05.014.

[80] Jordon JB. The effect of microstructural and geometrical features on fatigue performance in MG AZ31 friction stir spot welds. ASME 2011 International Mechanical Engineering Congress and Exposition, IMECE 2011, vol. 8; 2011.

[81] McDowell D, Gall K, Horstemeyer M, Fan J. Microstructure-based fatigue modeling of cast A356-T6 alloy. Eng Fract Mech 2003;70:49−80. Available from: https://doi.org/10.1016/S0013-7944(02)00021-8.

[82] Jordon JB, Gibson JB, Horstemeyer MF, Kadiri HEL, Baird JC, Luo AA. Effect of twinning, slip, and inclusions on the fatigue anisotropy of extrusion-textured AZ61 magnesium alloy. Mater Sci Eng A 2011;528:6860−71. Available from: https://doi.org/10.1016/j.msea.2011.05.047.

[83] Jordon JB, Gibson JB, Horstemeyer MF. Experiments and Modeling of Fatigue of an Extruded Mg AZ61 Alloy. Magnes. Technol. 2011. 2011. p. 55−60. Available from: https://doi.org/10.1002/9781118062029.ch13.

[84] Rettberg LH, Jordon JB, Horstemeyer MF, Jones JW. Low-cycle fatigue behavior of die-cast Mg alloys AZ91 and AM60. Metall Mater Trans A 2012;43:2260−74. Available from: https://doi.org/10.1007/s11661-012-1114-8.

[85] Lugo M, Jordon JB, Solanki KN, Hector LG, Bernard JD, Luo AA, et al. Role of different material processing methods on the fatigue behavior of an AZ31 magnesium alloy. Int J Fatigue 2013;52:131−43. Available from: https://doi.org/10.1016/j.ijfatigue.2013.02.017.

[86] Xue Y, McDowell DL, Horstemeyer MF, Dale MH, Jordon JB. Microstructure-based multistage fatigue modeling of aluminum alloy 7075-T651. Eng Fract Mech 2007;74:2810−23. Available from: https://doi.org/10.1016/j.engfracmech.2006.12.031.

[87] Allison PGG, Hammi Y, Jordon JBB, Horstemeyer MFF. Modelling and experimental study of fatigue of powder metal steel (FC-0205). Powder Metall 2013;56:388−96. Available from: https://doi.org/10.1179/1743290113Y.0000000063.

[88] Rodriguez RI, Jordon JB, Allison PG, Rushing TW, Garcia L. Corrosion effects on the fatigue behavior of dissimilar friction stir welding of high-strength aluminum alloys. Mater Sci Eng A 2019;742:255−68. Available from: https://doi.org/10.1016/j.msea.2018.11.020.

[89] Rao HM, Jordon JB. A multi-stage approach for predicting fatigue damage in friction stir spot welded joints of Mg AZ31 alloy. Magnes. Technol. 2013;197–222.

[90] McCullough RR, Jordon JB, Allison PG, Rushing T, Garcia L. Fatigue crack nucleation and small crack growth in an extruded 6061 aluminum alloy. Int J Fatigue 2019;119:52–61. Available from: https://doi.org/10.1016/j.ijfatigue.2018.09.023.

[91] Xue Y, El Kadiri H, Horstemeyer MF, Jordon JB, Weiland H. Micromechanisms of multi-stage fatigue crack growth in a high-strength aluminum alloy. Acta Mater 2007;55:1975–84. Available from: https://doi.org/10.1016/j.actamat.2006.11.009.

[92] Rao HM, Jordon JB, Barkey ME, Guo YB, Su X, Badarinarayan H. Influence of structural integrity on fatigue behavior of friction stir spot welded AZ31 Mg alloy. Mater Sci Eng A 2013;564:369–80. Available from: https://doi.org/10.1016/j.msea.2012.11.076.

CHAPTER 6

Extreme Conditions and Environments

6.1 INTRODUCTION

Extreme loading conditions and extreme environments tend to exacerbate the effects of fatigue. In the simplest of views, fatigue crack initiation and fatigue crack growth result from energy input into a material system. When a critical amount of energy is absorbed by the material, a crack may initiate (if one does not already exist) or an existing crack may grow. For the sake of generality, this chapter will refer to the manifestation of input energy resulting in crack initiation and crack growth as "damage." The energy resulting in damage may come from an external mechanical source (displacements or forces), electrical source, chemical source, or others. The amount of damage resulting from the energy input to the material system is a function of the magnitude, waveform, and frequency of the external energy source. Furthermore, the manifestation of mechanical energy may be increased locally as a function of stress triaxiality. Unfortunately, the amount of damage occurring within the material is rarely a result of the sum of all individual input energies. Rather, many of the input energies interact constructively with other energy types. That is to say, mechanical energy input to a system may cause dislocation motion. Oxygen may absorb and diffuse within this same material. The existence of oxygen within the material may result in grain boundaries with a reduced cohesive strength, among other factors. The oxygen may also interact with dislocation movement, thereby affecting the damage resulting from mechanical energy. The overall damage results from not only the individual components of input energy but also the interaction of these mechanisms with each other. When you couple the input of extreme loading conditions with a material having a graded microstructure, such as that produced as a result of friction stir welding (FSW), the effect may be altered yet again. Given a nonhomogeneous microstructure, the local energy state of the material as a function of weld zones

Fatigue in Friction Stir Welding. DOI: https://doi.org/10.1016/B978-0-12-816131-9.00006-4
© 2019 Elsevier Inc. All rights reserved.

and the gradient in the internal energy state will affect the overall damage accumulation. Specifically, large-grained microstructures may diffuse more oxygen than small grained ones, resulting in an increase in damage. Additionally, the gradient in energy state occurring between the FSW heat affected zone (HAZ) and nugget often result in stress concentrations as a function of deformation compatibility. In which case, multiaxial loading may result in failure at this location even in the absence of a geometric discontinuity. This chapter will provide background on some key fatigue mechanisms that result in a modified response in materials including variable amplitude fatigue, multiaxial fatigue, corrosion fatigue, and effect of prestraining. These fatigue mechanisms represent different realistic loading and boundary conditions that modify the energy state of the material, often resulting in increased rates of damage accumulation. As the FSW process is relatively new, and given that extreme loading conditions are relatively onerous, one will note that the list of publications provided is somewhat limited.

6.2 VARIABLE AMPLITUDE FATIGUE

Variable amplitude loading occurs when a fatigue waveform is not consistent with respect to time. While in-service loading conditions are rarely consistent with respect to time, the vast majority of the laboratory testing is performed by use of a simple sine-wave or saw-tooth wave loading function. The flight spectrum applied to aero turbine components is an example of realistic in-service variable amplitude loading. An aero turbine flight spectrum consists of starting the turbine, taxying of the aircraft, takeoff, flight, landing, taxying, and shut down. During flight the civil aero turbine may exhibit a generally consistent load spectrum, but it also may exhibit load spikes resulting from maneuvers. A generic flight spectrum is provided in Fig. 6.1

Note that the takeoff and landing portions of the flight spectrum may impart damage similar to that found in simple low-cycle fatigue (LCF) loading, whereas the in-flight spectrum may impart high-cycle fatigue (HCF)-like damage. The differences in energy input and resulting damage may interact constructively or destructively, depending upon the order in which they are encountered. As such, variable amplitude loading is one of the more challenging loading conditions for an engineer to design for.

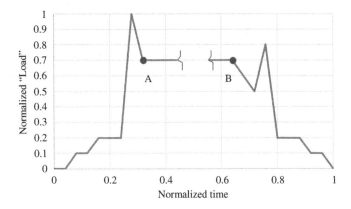

Figure 6.1 Generic aero flight loading spectrum. Note that the locations in the flight spectrum designated "A" and "B" demarcate in flight conditions. Though the figure indicates a constant load with respect to time, fatigue loading will occur within this region.

One typically designs for variable amplitude loading by use of a damage summation rule. Damage summation rules state that failure will occur as a result of a critical accumulation of damage. If the damage is normalized, the damage rule estimates failure when

$$\sum D_i \geq 1,$$ (6.1)

where D_i is the damage accrued from the i-th load scenario. If the external loading can be separated into primary damage events, or perhaps the variable amplitude loading can be discretized into like-events, one can use Miner's rule to determine the amount of damage from the i-th load scenario. Specifically, the amount of damage resulting from the i-th load scenario is defined as

$$D_i = \frac{n_i}{N_i},$$ (6.2)

where n_i is the number of cycles accrued during load scenario i, associated with a nominal stress σ_i, and N_i is the number of cycles to failure for the nominal stress σ_i. An example of the use of the Minor's rule in predicting variable amplitude loading in FSW is shown in Fig. 6.2 [1]. In this work, Costa et al. [1] carried out variable amplitude fatigue tests on AA6082 butt-friction stir welding (FSBW) and observed acceptable correlation using Miner's rule for $R = 0$, but significantly more unconservative results were achieved for fully reversed loading ($R = -1$). In addition, Costa et al. assessed the use of the double linear damage rule by Manson and Halford [2], where the linear damage rule

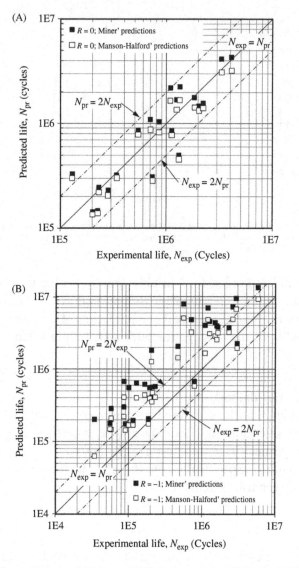

Figure 6.2 Example use of Miner's and Manson and Halford rules applied to (A) R = 0 and (B) R = −1 experimental variable amplitude fatigue results for AA6082 FSBW joints [1].

is applied twice, and found that this approach produced slightly more conservative predictions for both $R = 0$ and -1.

If damage accumulation is not linear with respect to different loading scenarios, one could also use a nonlinear damage rule

$$D_i = \left(\frac{n_i}{N_i}\right)^k, \tag{6.3}$$

where k is a function of many factors, including stress level. The difficulty of employing a damage summation law for variable amplitude loading arises when one attempts to differentiate "like" load scenarios and their associated nominal stresses. Using the flight load spectrum as an example, perhaps the LCF-like loads imparted by the takeoff and landing would constitute a single load scenario, per flight. The HCF-like loads imparted during flight may then constitute a separate load scenario, per flight. One would then determine the sum total of damage by the use of

$$D_{total} = \frac{n_{LCF}}{N_{\sigma_{LCF}}} + \frac{n_{HCF}}{N_{\sigma_{HCF}}}, \tag{6.4}$$

where LCF in this case refers to the takeoff and landing portion and HCF refers to the in-flight portion of the variable amplitude loading. The stresses σ_{LCF} and σ_{HCF} are the stresses associated with the takeoff/landing and in-flight portion of the variable amplitude loading, respectively. Implementation of an LCF and HCF linear damage rule for FSW was demonstrated by Joy-A-Ka et al. [3] for a two-step force amplitude. However, they found that this modified Miner's rule resulted in a nonconservative prediction. Joy-A-Ka et al. reported more conservative predictions when using a modified regression approach suggested by Haibach [4].

There are many methods to determine load incursions that create similar damage that may therefore be lumped into a single load scenario for damage summation. These methods include level-crossing, peak counting, and rainflow counting. In general, these methods are referred to as cycle counting techniques. Ultimately, these methods combine "like" sections of the variable amplitude load spectrum such that the combined subcycles may be compared to a single laboratory constant amplitude test to calculate $\frac{n_i}{N_i}$ for that set of load incursions. By use of one of these methods, the n_i's and σ_i's may be determined, compared to the associated N_i for each σ_i, and an estimate of accumulated damage may be determined. The cycle counting techniques are beyond the scope of this book and the interested reader is referred to a fatigue text such as Ref. [5].

As mentioned in the introduction, the microstructures manifested as a function of the FSW process will accumulate damage differently from one another as a function of their respective responses to external mechanical loads. The FSW stir zone, as example, may have a decreased yield stress as a result of the heat input to the base material or may have an increased yield stress as a function of grain refinement. Nontheless, the yield stress in the stir zone is likely different from that produced in the thermo mechanically affected zone (TMAZ). As such, damage will accumulate differently in different regions of the FSW and base material and likely require specially modified variable models to address nonconservative predictions similar to the results reported by Costa et al. [1].

Fracture mechanics concepts can also be used to predict variable amplitude by accounting for the mean stress and crack closure effect described in Chapter 4, Fatigue Crack Growth in Friction Stir Welds. In fact, it is well established that variation in load sequences can lead to significantly different crack growth rates [6,7]. Implementation of such load sequences effects have been implemented in software packages developed for predicting the structural integrity in aging aircraft like AFGROW [8] and NASGROW [9]. An example of the use of the damage-tolerant approach with load sequence effects in predicting variable amplitude loading in FSW joints is presented in work by Ghidini and Donne [10]. As shown in Fig. 6.3, Ghidini and Donne

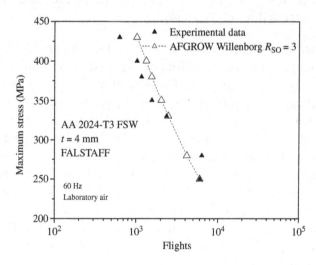

Figure 6.3 An example of using crack growth retardation model to predict variable amplitude loading on FSW coupons with an open hole [10].

demonstrated good results for predicting AA2024 FSW joints subjected to aircraft flight loading history [11] by using a crack retardation model by Willenborg et al. [12] that is implemented in the AFGROW software. In the Willenborg model, the stress intensity factor is modified by using an effective stress concept. Other load interaction models, including the ones implemented in the widely used software packages like AFGROW and NASGROW, require further evaluation to determine their effectiveness in predicting variable amplitude loading for structures joined by FSW.

6.3 MULTIAXIAL FATIGUE

Multiaxial fatigue is a general term that may be used to describe loading and/or loading plus geometry conditions resulting in complex states of stresses and strains, either locally or globally. More specifically, multiaxial loading will result in a state of stress and/or strain, which manifests as two or more components in the stress or strain tensor. An example of a stress tensor resulting at the surface of a component experiencing externally applied normal force, coupled with an externally applied toque, is provided below.

$$\sigma_{ij} = \begin{bmatrix} \sigma & \tau \\ \tau & 0 \end{bmatrix}$$

Note that the multiaxial aspect of the above example resulted from separate external loads (forces and torques) being applied. As such, this is an example of global multiaxial loading. Global multiaxial loading may also result from multiple external forces being applied along different lines of action. Local multiaxial loading, on the other hand, will occur at any geometric discontinuity, even when the global state of stress is uniaxial. A particular note for designers is that all notches, holes, bends in structures, and so on, will experience multiaxial loading, and therefore multiaxial fatigue.

The state of stress within a component experiencing multiaxial loading is best determined by the use of finite element methods and is therefore somewhat trivial. Failure criterion for multiaxial fatigue, on the other hand, is the subject of ongoing research and remains a difficult task for engineers and scientists. One failure criterion that has been successfully applied to multiple metals, loading conditions, and boundary conditions is the Fatemi–Socie (F-S) fatigue indicator parameter (FIP) [13]. Given the popularity of the F-S FIP, calibration constants for many common materials are available in the literature. The F-S FIP estimates that there will be failure when a critical combination of shear strain and normal stress occur on a critical plane. The F-S FIP equation is provided below.

$$\text{FIP}_{\text{F-S}} = \left(\frac{\Delta\gamma}{2} \times \left[1 + K\left(\frac{\sigma_{norm\_max}}{\sigma_y} \right) \right] \right) \qquad (6.5)$$

The term $\Delta\gamma$ is the shear strain range on the plane of interest, K is a material constant that must be determined from experiment, σ_{norm_max} is the maximum normal stress on the plane of interest, and σ_y is the material yield stress. K has been suggested to be a value of 0.5 with a maximum value of 1. Note that in order to determine the F-S FIP, one must know the plane of interest (e.g., close-packed crystallographic plane) or one must sample planes to determine the parameter.

A similar approach to predicting multiaxial fatigue is the use of the Modified Wöhler Curve Method (MWCM). This method, based on the concept of a critical plane, uses the assumption that damage from fatigue is at a maximum on planes of corresponding maximum shear stress amplitudes [14]. In the study by Susmel et al. [14], postmortem examination of FSW AA6082 tubes subjected to stress-controlled multiaxial fatigue loading of various combinations of axial and shear stress revealed that fatigue crack incubation was observed to be shear stress dominated. As such, the MWCM approach appears to be appropriate for FSW of aluminum tubes, where the model correlation results fell within the upper and lower bounds of uniaxial and torsional experiments. Fig. 6.4 illustrates the use of the MWCM for modeling the multiaxial fatigue loading in FSW of aluminum tubes.

We conclude the discussion of multiaxial fatigue by acknowledging that determining the effects of multiaxial fatigue is further complicated by the existence of a graded microstructure in FSW. First, the transition in microstructure between the stir zone and TMAZ, TMAZ and HAZ, and HAZ and base material have a propensity to manifest as local stress concentrations. That is, localized portions of these regions are likely to exhibit a multiaxial state of stress even when the long-range state of stress is uniaxial. Further exacerbating the situation, the TMAZ region surrounding an FSW typically presents a textured grain morphology. As such, the alignment of the critical planes within the

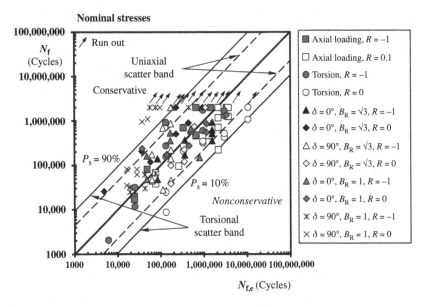

Figure 6.4 Example of use the Modified Wöhler Curve Method for modeling multiaxial fatigue loading in FSW [14].

TMAZ is not random, and likelihood of favorable alignment within regions of stress triaxiality may manifest in multiaxial fatigue-like damage accumulation. Thus, future work is needed to develop multiaxial approaches that consider the microstructure in FSW.

6.4 CORROSION FATIGUE

Corrosion fatigue is a scenario in which damage of a material occurs as a result of cyclic loading in a corrosive medium. Note that, as a result of electrochemical reactions, seemingly innocuous medium such as water are also corrosive to many metals. In fact, corrosion fatigue studies in water or an aqueous medium are likely the most common in the literature. While there are many species that contribute to corrosion fatigue, a concise outline of the corrosion fatigue process in water is provided below.

Corrosion fatigue in water results from electrochemical reactions at the surface of the material. These electrochemical reactions will experience higher reaction rates at locations on the material in which there is a gradient in the local energy state (e.g., pit, slip-step to surface interface, crack tip, etc.). Once water is at one of these locations, one of several reactions may occur. A likely reaction results in the dissociation of oxygen and hydrogen from the water molecules. While the oxygen is involved in the reduction−oxidation (re-dox) reaction, the hydrogen may be adsorbed and subsequently absorbed into the metal. Of note here is that corrosion fatigue may manifest on the surface of the material, as in the re-dox reaction discussed or within the bulk of the material as in hydrogen absorption. The effect of atomic hydrogen in metals is an ongoing research area and will not be discussed here.

Corrosion fatigue in a gaseous medium also occurs as a result of re-dox reactions occurring at the surface of the material. High-temperature fatigue and thermo-mechanical fatigue are two examples of corrosion fatigue that occur in air. The re-dox reaction occurring at the surface of the material creates an oxide as well as dissociates oxygen atoms from the air molecule. The dissociation of the oxygen atom enables the uptake of oxygen into the bulk of the material. The oxygen atom within the material may create subsurface oxides, decrease the cohesion of grain boundaries, and hinder dislocation movement, among other things.

The rate controlling steps for corrosion fatigue include the rate of passivation layer formation (if any), the rate of passivation layer rupture (typically resulting from external fatigue loading), the availability of corrosive species (e.g., water, oxygen, etc.), the rate of species adsorption and subsequent absorption, and the rate of corrosive species' diffusion within the material if concern is subsurface. Ultimately, corrosion damage and fatigue damage are rarely additive; as such failure modeling of corrosion fatigue is a nontrivial task for homogeneous microstructures. The differential in energy state of the microstructures resulting from the FSW process, and especially the interfaces between the microstructural regions, are prime high-energy regions for species diffusion. As such, FSW joints will experience corrosion fatigue at rates considerably different than those found in the base material.

Milan et al. [15] examined FGC rates for FSW under a salt fog environment and in air for cracks orientated longitudinally within the weld zone and for cracks growing perpendicular to the weld. As shown in Fig. 6.5, for fatigue cracks orientated longitudinally within the weld zone, they found that FCG rates were slower when subjected to the salt fog then when compared to the fatigue cracks that grew in air for low ΔK. This slower FCG rate in the salt fog environment was due to

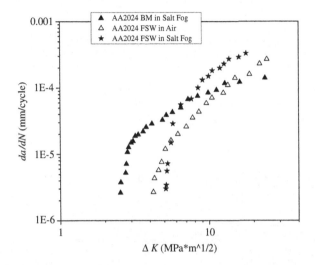

Figure 6.5 Fatigue crack growth rates for cracks growing in the weld zone in a salt fog and in air. Data replotted from Milan MT, Bose WW, Tarpani JR. Fatigue crack growth behavior of friction stir welded 2024-T3 aluminum alloy tested under accelerated salt fog exposure, Mater Perform Charact 3 (2014) 20130036. doi:10.1520/MPC20130036.

crack closure produced by corrosion debris. In contrast, similar FCG rates were observed for both corrosive and air environments. This observation was likely due to competing mechanisms, where the accelerated FCG rates because of hydrogen embrittlement was opposed by FCG retardation from crack closure and compressive residual stress fields. For intermediate ranges of ΔK, the effect of hydrogen embrittlement due to the salt fog resulted in FCG rates that were up to five times greater when compared to in air.

In addition to measuring FCG rates of FSW in corrosive environments, several studies have investigated the effect of precorrosion on fatigue behavior [10,16]. In particular, Rodriguez et al. [16] examined the effects of precorrosion on the strain-life performance of dissimilar FSBW AA6061-to-AA7050. In this study, corrosion pitting was produced on the crown surface of the weld by static immersion in 3.5% NaCl for various exposure times. Fig. 6.6 shows the fatigue specimen used in their study, where the specimens were masked except for the crown surface of the FSBW. Results from their study revealed localized corrosion damage in the TMAZ and HAZ of the dissimilar FSW joints, as shown in Fig. 6.7. In particular, the corrosion damage featured general pitting, pit clustering, and exfoliation that revealed increasing depth with increasing exposure time. The higher corrosion

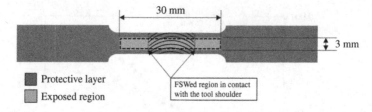

Figure 6.6 Schematic representation of the sample preparation for the precorroded fatigue test [16].

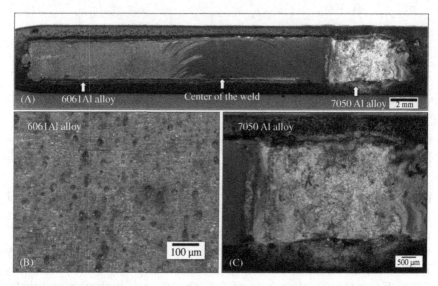

Figure 6.7 (A) Corrosion damage produced in a dissimilar FSBW of AA6061-to-AA7050 after 30 days immersion in 3.5%-NaCl aqueous solution. (B) Pitting damage in the AA6061 side. (C) Severe corrosion damage extending from the beginning of the TMAZ towards the AA7050 base material [16].

attack was measured on the AA7050 side of the weld compared to the AA6061 side.

After producing FSBW joints with varying corrosion pits, strain-life fatigue experiments were carried out. Fig. 6.8 shows the effect of pre-corrosion damage on strain-life for the dissimilar AA6061-to-AA7050 FSW specimens. Also shown in Fig. 6.8 are the effect of precorrosion damage on similar AA6061 and AA7050 FSBW joints. The experimental fatigue results demonstrated a decrease in the strain-life for the dissimilar AA6061-to-AA7050 and the similar AA7050 FSBW joints with evidence of fatigue crack incubation at surface corrosion defects. Regarding the dissimilar AA6061-to-AA7050 FSBW specimens, Rodriguez et al. observed crack propagation and fracture on the

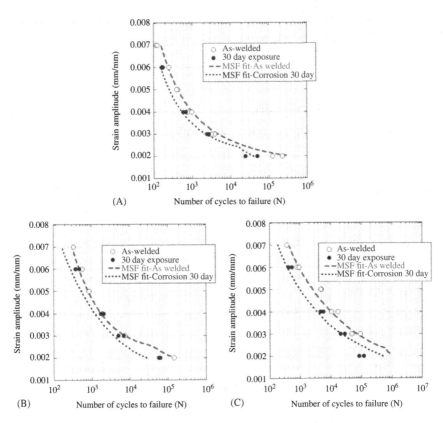

Figure 6.8 Comparison of strain-life experiments to the MSF model for the FSBW (A) AA6061-to-AA7050, (B) AA6061-to-AA6061, and (C) AA7050-to-AA7050 in the as-welded and precorroded condition [16].

AA6061 side of the weld at strain amplitudes greater than 0.3% and crack propagation and fracture in the AA7050 side of the weld at strain amplitudes below 0.3%. They conclude that the corrosion damage appeared to shift the HCF mechanisms from occurring on the AA6061 side of the weld (due to the lower general strength) to the AA7050 side of the weld, where the dominant fatigue crack nucleated due to the local stress concentration created by the corrosion damage. In addition, the microstructure-sensitive fatigue (MSF) model presented in Chapter 5, Fatigue Modeling of Friction Stir Welding, was used to correlate the fatigue life due to corrosion pitting. As such, good correlation between the MSF model and the precorrosion experiments were observed. In particular, Rodriguez et al. showed that the MSF model captured the effect of precorrosion on fatigue behavior by treating the defects as inclusions within the existing modeling structure.

As such, the multistage framework of the MSF model appears robust enough to capture the effect precorroded damage on fatigue behavior observed in FSW.

6.5 EFFECT OF PRESTRAIN

In this last section, we discuss the effect of prestraining on fatigue behavior. While the need to investigate the effect of prestrain prior damage on fatigue performance may not be obvious, one should note that prestrain may occur due to in-service overloads or as part of the manufacturing process. An example of prestrain resulting from the manufacturing process is tailor weld blanks. Tailor weld blanks are made from welded sheets that are then stamped to shape, resulting in significant plastic deformation of FSW before any cyclic loading. Recent work by White et al. [17] examined the plastic strain gradients developed by prestraining AA7050 FSBW joints, the damage to intermetallic particles caused by this strain, and the effect of these fractured particles on the subsequent fatigue performance of the weld. In this study, AA7050 FSBW were subjected to monotonic strains of 0%, 1%, and 3%, where the local plastic strain distribution was obtained by measuring the spacing between microhardness indentations before and after prestraining.

Prestraining FSBW is complicated by the strength gradient in the FSBW joint, which causes a plastic strain gradient to form when a mechanical load is applied, as shown in Fig. 6.9. A number of interesting observations can be made from the plastic strain maps, which relate not only to prestraining but also to the nature of the mechanical properties of FSBW joints in general. More importantly, it is clear that plastic strain is highly localized to the HAZ, and in fact, the stir zone (SZ) essentially experiences no deformation at all after 1% strain. This is apparent from quantitative strain measurements but can also be seen qualitatively from the visual appearance of the prestrained welds (Fig. 6.9B–E), where surface deformation bands can be seen in the HAZ. The 3% prestrain map shows that not only is the retreating side weaker than the advancing side, but there is a strain (and therefore strength) gradient within the SZ. Finally, by comparing the strains on both the top and bottom of the weld at any given location, it is evident that one side of the weld has plastically strained more than the other, which results in bending of the FSBW joint, even under uniaxial tensile loading.

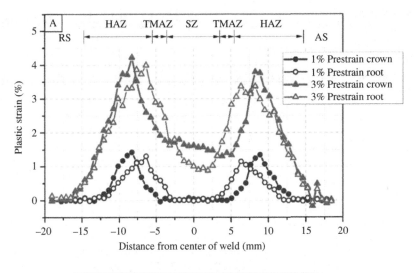

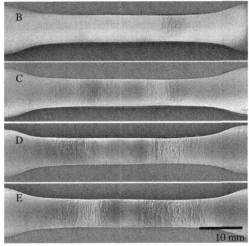

Figure 6.9 (A) Plastic strain maps after 1% and 3% prestrain for an AA7050 FSBW. Visual appearance of polished welds after prestraining: (B) 1% prestrain crown, (C) 1% prestrain root, (D) 3% prestrain crown, (E) 3% prestrain root [17].

Metallography representation of the cross-section of the FSBW joint, presented in Fig. 6.10, shows that the fraction of cracked Mg_2Si and Fe-rich particles increased with increasing prestrain. Furthermore, both Mg_2Si and Fe-rich particles were found at the fatigue initiation sites, with the 3% group showing Fe-Rich particles at majority of the fatigue initiation sites.

The strain-life curves in Fig. 6.11 show that only the 1% prestrain condition had any effect on the fatigue life and even then only at lower

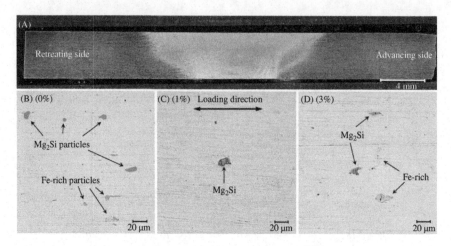

Figure 6.10 (A) Cross-sectional view of a AA7050 FSBW. Micrographs showing progressively more cracked particles with increasing prestrain at (B) 0%, (C) 1%, and (D) 3% prestrain. [17].

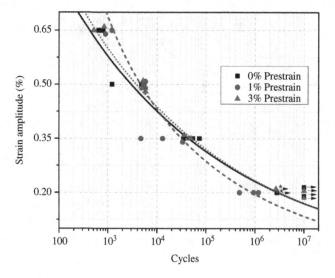

Figure 6.11 Strain-life curves for AA7050 FSW after 0%, 1%, and 3% prestrain [17].

strain amplitudes. The observation that the 3% prestrain condition did not show the same detrimental effect was most likely because most of the damage to intermetallic particles had already occurred by 1% prestrain, and further prestraining served only to grow the plastic zone around these cracked particles and thus work harden the material, leading to improved fatigue resistance. Unlike wrought AA7050 which has a relatively flat monotonic stress—strain response, the FSBW joints exhibited a 27% increase in yield strength after prestraining to 3%, which may explain why the detrimental prestrain effect is observed in wrought materials but disappears at higher prestrains in the FSBW joints.

REFERENCES

[1] Costa JD, Ferreira JAM, Borrego LP, Abreu LP. Fatigue behaviour of AA6082 friction stir welds under variable loadings. Int J Fatigue 2012;37:8—16. Available from: https://doi.org/ 10.1016/j.ijfatigue.2011.10.001.

[2] Manson SS, Halford GR. Practical implementation of the double linear damage rule and damage curve approach for treating cumulative fatigue damage. Int J Fract 1981;17:169—92.

[3] Joy-A-Ka S, Ogawa Y, Akebono H, Kato M, Sugeta A, Sun Y, et al. Fatigue damage evaluation of friction stir spot welded cross-tension joints under repeated two-step force amplitudes. J Mater Eng Perform 2015;24:2494—502. Available from: https://doi.org/10.1007/ s11665-015-1534-x.

[4] Haibach E. The allowable stresses under variable amplitude loading of welded joints. In: Proceeding Conference Fatigue of Welded Structure, The Welding Institute, 1971. pp. 328—39.

[5] Bannantine JA, Comer JJ, Handrock JL. Fundamentals of Metal Fatigue Analysis. Englewood Cliffs, NJ: Prentice Hall; 1990.

[6] Skorupa M. Load interaction effects during fatigue crack growth under variable amplitude loading—a literature review. Part I: empirical trends. Fatigue Fract Eng Mater Struct 1998;21(8):987—1006. Available from: https://doi.org/10.1046/j.1460-2695.1998.00083.x.

[7] Skorupa M. Load interaction effects during fatigue crack growth under variable amplitude loading—a literature review. Part II: qualitative interpretation. Fatigue Fract Eng Mater Struct 1999;22:905—26. Available from: https://doi.org/10.1046/j.1460-2695.1999.00158.x.

[8] Harter JA. AFGROW users guide and technical manual. Air Force Research Lab Wright-Patterson AFB OH Air Vehicles Directorate, 1999.

[9] Mettu SR, Shivakumar V, Beek JM, Yeh F, Williams LC, Forman RG, et al., NASGRO 3.0: A software for analyzing aging aircraft. 1999.

[10] Ghidini T, Dalle Donne C. Fatigue life predictions using fracture mechanics methods. Eng Fract Mech 2009;76:134—48. Available from: https://doi.org/10.1016/j.engfracmech.2008.07.008.

[11] FALSTAFF A. Description of a Fighter Aircraft Loading STAndard for Fatigue Evaluation. Switzerland: F + W; The Netherlands: LBF IABG NLR; 1976.

[12] Willenborg J, Engle RM, Wood HA. A crack growth retardation model using an effective stress concept. Tech Memo 1971; 71-1-FBR.

[13] Fatemi A, Socie DF. A critical plane approach to multiaxial fatigue damage including out-of-phase loading. Fatigue Fract Eng Mater Struct 1988;11:149—65.

[14] Susmel L, Hattingh DG, James MN, Tovo R. Multiaxial fatigue assessment of friction stir welded tubular joints of Al 6082-T6. Int J Fatigue 2017;101:282—96. Available from: https:// doi.org/10.1016/j.ijfatigue.2016.08.010.

[15] Milan MT, Bose WW, Tarpani JR. Fatigue crack growth behavior of friction stir welded 2024-T3 aluminum alloy tested under accelerated salt fog exposure. Mater Perform Charact 2014;3:20130036. Available from: https://doi.org/10.1520/MPC20130036.

[16] Rodriguez RI, Jordon JB, Allison PG, Rushing TW, Garcia L. Corrosion effects on fatigue behavior of dissimilar friction stir welding of high-strength aluminum alloys. Mater Sci Eng A 2019;742:255–68. Available from: https://doi.org/10.1016/j.msea.2018.11.020.

[17] White BC, White RE, Jordon JB, Allison PG, Rushing T, Garcia L. The effect of tensile pre-straining on fatigue crack initiation mechanisms and mechanical behavior of AA7050 friction stir welds. Mater Sci Eng A 2018;736:228–38. Available from: https://doi.org/10.1016/j.msea.2018.08.104.

CHAPTER 7

Beyond Friction Stir Welding: Friction Stir Processing and Additive Manufacturing

7.1 INTRODUCTION

While this book focuses on fatigue in friction stir welding (FSW), it is pertinent to briefly discuss several related topics. The first related topic to discuss is friction stir processing (FSP), which is a derivative of FSW. FSP involves essentially the same physics that occur in FSW, except that in FSP, the main purpose is to process the material by refining the microstructure and not join separate materials together as in FSW. The goal of FSP is to process materials with multiple passes of the weld tool on a substrate such that the desired area has been fully "stirred" with the weld tool. As such, FSP will generally produce a refined microstructure over a much larger area than the typical narrow weld zone in FSW. The other topic of interest to discuss briefly in this chapter is a relatively new additive manufacturing (AM) process based on similar physics to that of FSW. This novel AM process, called additive friction stir deposition (AFS-D), uses high-shear deformation to achieve solid-state depositions that are fully dense, near-net shaped components with superior mechanical properties.

7.2 FRICTION STIR PROCESSING

The attractiveness in using FSP is to take advantage of the beneficial properties of the weld zone while mostly doing away with the thermo mechanically affected zone and heat affected zone. Studies have shown that FSP of certain materials can have a considerable positive impact on fatigue performance. In a study by Jana et al. [1], FSP of cast A357 aluminum alloy exhibited up to 5 times improvement in the number of cycles to failure in the low cycle fatigue regime compared to the unprocessed cast material, as shown in Fig. 7.1. It is important to point out that their study analyzed both full thicknesses processed and only partially processed samples, where both types of processes exhibited significantly

Fatigue in Friction Stir Welding. DOI: https://doi.org/10.1016/B978-0-12-816131-9.00007-6
© 2019 Elsevier Inc. All rights reserved.

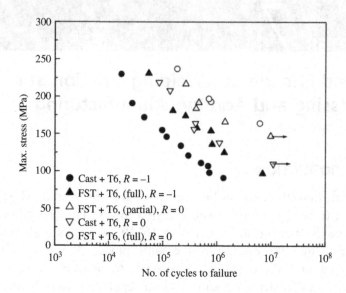

Figure 7.1 Comparison of S–N plot of friction stir processed samples and cast aluminum alloy samples at R = 0 and −1 [1].

better stress-life than the cast material. Even more impressive is when the FSP samples were tested at stress levels near run-out levels (high cycle fatigue), the improvement in fatigue life increased by a factor of 15. This improvement was attributed to refinement of the silicon particles, reduction of porosity, and increased level of crack closure in the FSP samples compared to the cast samples. While not all literature data suggest similarly high improvement in fatigue performance as the study by Jana et al. [1], studies have reported general improvement in fatigue performance from FSP due to the refinement of the microstructure and reduction of casting voids [2–6]. It is also important to point out that FSP of wrought materials may not show much improvement and even a reduction in fatigue performance. Since most rolled aluminum alloys have little or no casting pores, and no detrimental dendritic microstructures, these rolled materials are very near their optimized fatigue strength after precipitation heat treatment. After FSP, the added heat input would likely undo heat treatment by causing some degree of dissolution of the precipitates, and thus it is expected that the FSP would likely not lead to any improvement in fatigue properties of these wrought materials. In fact, Alfrinaldi et al [7] showed that the fatigue life of FSP AMX602 samples exhibited lower fatigue lives compared to the wrought samples. The gap in the fatigue lives between the FSP and the wrought samples was closed, but not completely, after post-aging of the FSP samples.

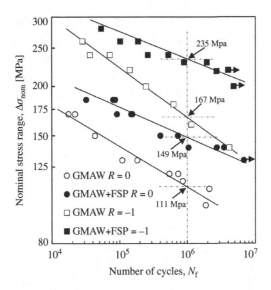

Figure 7.2 Example of improvement in fatigue behavior of gas metal arc welding (GMAW) after friction stir welding (FSW) for 5083 aluminum alloys [8].

The benefit of FSP is not just limited to cast materials. FSP can also be used to improve fusion welds by refining the microstructure and reducing or eliminating welding defects. In fact, the improvement in fatigue performance of fusion welds modified using FSP may exhibit as much improvement as for the cast materials. Thus, the FSP represents a unique and potential method for improving the mechanical properties of traditional fusion welds. The work by Jesus et al. [8] revealed that using FSP on gas metal arc welded butt joints of 5083 aluminum alloy improved the fatigue strength at 10^6 cycles by 41%, as shown in Fig. 7.2. They concluded that the improvement in the fatigue performance was mainly attributed to the ablation of welding defects such as porosity and lack of wetting, with a minor contribution from the refinement of the grains. Additional studies using FSP to improve the fatigue properties of fusion welds report similar findings [9–12].

7.3 ADDITIVE FRICTION STIR DEPOSITION

The AFS-D process, commercially known as MELD, is a solid-state AM technology that uses either a solid rod or powder to produce fully dense components for a wide range of materials. In the MELD process, the feedstock materials are fed through a nonconsumable rotating cylindrical tool generating heat and plastically deforming the feedstock material through a controlled pressure as sequential layers are built on each other, as shown in Fig. 7.3A. Previous studies have shown that this process can produce a refined microstructure resulting in higher mechanical properties compared to other fusion-based AM processes [13,14].

A recent study by Avery et al. [15] focused on the characterizing the fatigue behavior of Inconel 625 processed with MELD. Fig. 7.3B shows an example of the MELD as-deposited Inconel 625 deposited on the HY80 substrate. The feedstock material used in this study was an extruded hot rolled and annealed ATI IN625. For this work, a modified sub-scaled hourglass fatigue coupon was adopted, shown in Fig. 7.3C. The MELD process can, because of the high-shear deformation process, dramatically refine the grain structure. As such, the as-deposited IN625 MELD exhibited an average grain size reduction from 30 μm in the feedstock to 1 μm. Additionally the interfacial layer was observered to exhibited a sub-micron average grain size of 0.26 μm. The refined grain size in the as-deposited material is a product of the dynamic recrystallization of the grains due to high-shear deformation [14].

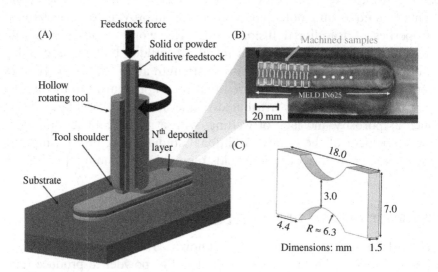

Figure 7.3 (A) Schematic of high-shear deposition process. (B) As-deposited IN625. (C) Fatigue specimen geometry [15]. Reprinted with permission from Springer Nature.

Since the focus of this study was on the behavior of fatigue in the MELD process, Avery et al. [15] also characterized other aspects of the microstructure including carbide particles. Fig. 7.4A and B display the optical micrographs of both feedstock and as-deposited IN625, respectively, where carbides $M_{23}C_6$ and M_6C [16] were seen distributed throughout the matrix. Fig. 7.4C shows the normal distribution of inclusions based on the area of the particles. It is important to note that the as-deposited material exhibited a decrease in the average carbide size and had a narrower distribution of average particle sizes, indicative of uniformly dispersed constitutive particles. The lower average carbide size is suggestive of a more homogenous microstructure due to the mechanical processing of the high-shear deposition process.

Fig. 7.5 displays the stress-life experimental results of the load-controlled fatigue tests of the as-deposited IN625 and the corresponding feedstock material. As seen in Fig. 7.5, the as-deposited IN625 exhibited generally greater fatigue resistance compared to wrought

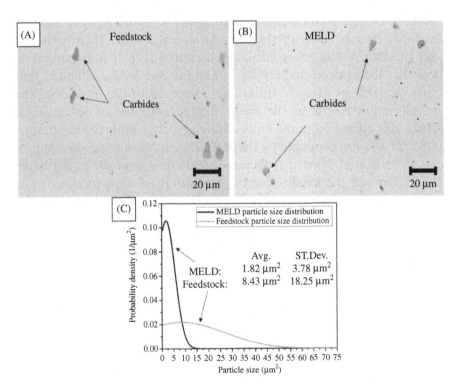

Figure 7.4 (A) Optical micrograph of particles in feedstock IN625. (B) Optical micrograph of particles in MELD IN625. (C) Particle size distribution of MELD IN625 and feedstock. [15]. Reprinted with permission from Springer Nature.

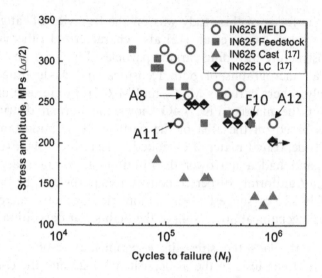

Figure 7.5 Stress-life fatigue results comparing as-deposited IN625 to the feedstock, cast, and laser consolidation materials [15]. Reprinted with permission from Springer Nature.

IN625 feedstock. Additionally, the as-deposited IN625 MELD material exhibited fatigue behavior that was on average better than both cast and LC IN625 data taken from the literature [17]. It is important to note that the as-deposited IN625, at certain load levels, exhibited four times the number of cycles to failure compared to the feedstock. The overall improvement in the mean fatigue behavior of as-deposited IN625 was attributed to the refined microstructure and refined intermetallic particle size produced by the high-shear solid-state deposition process. The aforementioned reduction in carbide size in the as-deposited IN625 lowered the localized stress concentration, thus increasing the number of cycles to incubate a fatigue crack as well as increasing the fatigue crack growth resistance, and ultimately improving the overall fatigue performance of the material. However, the fatigue life of as-deposited IN625 did exhibit some scatter suggesting that additional process optimization is needed. In summary, this fatigue study on a new AFS-D process, which shares many of the benefits of FSW, shows great promise in using solid-state high-shear deformation process to additively manufacture highly fatigue resistance materials.

REFERENCES

[1] Jana S, Mishra RS, Baumann JB, Grant G. Effect of stress ratio on the fatigue behavior of a friction stir processed cast Al-Si-Mg alloy. Scr Mater 2009;61:992−5. Available from: https://doi.org/10.1016/j.scriptamat.2009.08.011.

[2] Ni DR, Chen DL, Yang J, Ma ZY. Low cycle fatigue properties of friction stir welded joints of a semi-solid processed AZ91D magnesium alloy. Mater Des 2014;56:1–8. Available from: https://doi.org/10.1016/j.matdes.2013.10.081.

[3] Tajiri A, Uematsu Y, Kakiuchi T, Tozaki Y, Suzuki Y, Afrinaldi A. Effect of friction stir processing conditions on fatigue behavior and texture development in A356-T6 cast aluminum alloy. Int J Fatigue 2015;80:192–202. Available from: https://doi.org/10.1016/j.ijfatigue.2015.06.001.

[4] Uematsu Y, Tokaji K, Fujiwara K, Tozaki Y, Shibata H. Fatigue behaviour of cast magnesium alloy AZ91 microstructurally modified by friction stir processing. Fatigue Fract Eng Mater Struct 2009;32:541–51. Available from: https://doi.org/10.1111/j.1460-2695.2009.01358.x.

[5] Jana S, Mishra RS, Baumann JB, Grant G. Effect of friction stir processing on fatigue behavior of an investment cast Al-7Si-0.6 Mg alloy. Acta Mater 2010;58:989–1003. Available from: https://doi.org/10.1016/j.actamat.2009.10.015.

[6] Nelaturu P, Jana S, Mishra RS, Grant G, Carlson BE. Influence of friction stir processing on the room temperature fatigue cracking mechanisms of A356 aluminum alloy. Mater Sci Eng A 2018;716:165–78. Available from: https://doi.org/10.1016/j.msea.2018.01.044.

[7] Afrinaldi A, Kakiuchi T, Itoh R, Mizutani Y, Uematsu Y. The effect of friction stir processing and post-aging treatment on fatigue behavior of Ca-added flame-resistant magnesium alloy. Int J Adv Manuf Technol 2018;95:2379–91. Available from: https://doi.org/10.1007/s00170-017-1411-x.

[8] Jesus JS, Costa JM, Loureiro A, Ferreira JM. Fatigue strength improvement of GMAW T-welds in AA 5083 by friction-stir processing. Int J Fatigue 2017;97:124–34. Available from: https://doi.org/10.1016/j.ijfatigue.2016.12.034.

[9] da Silva J, Costa JM, Loureiro A, Ferreira JM. Fatigue behaviour of AA6082-T6 MIG welded butt joints improved by friction stir processing. Mater Des 2013;51:315–22. Available from: https://doi.org/10.1016/j.matdes.2013.04.026.

[10] Ito K, Okuda T, Ueji R, Fujii H, Shiga C. Increase of bending fatigue resistance for tungsten inert gas welded SS400 steel plates using friction stir processing. Mater Des 2014;61:275–80. Available from: https://doi.org/10.1016/j.matdes.2014.04.076.

[11] Borrego LP, Costa JD, Jesus JS, Loureiro AR, Ferreira JM. Fatigue life improvement by friction stir processing of 5083 aluminium alloy MIG butt welds. Theor Appl Fract Mech 2014;70:68–74. Available from: https://doi.org/10.1016/j.tafmec.2014.02.002.

[12] De Jesus JS, Loureiro A, Costa JM, Ferreira JM. Effect of tool geometry on friction stir processing and fatigue strength of MIG T welds on Al alloys. J Mater Process Technol 2014;214:2450–60. Available from: https://doi.org/10.1016/j.jmatprotec.2014.05.012.

[13] Rivera OG, Allison PG, Brewer LN, Rodriguez OL, Jordon JB, Liu T, et al. Influence of texture and grain refinement on the mechanical behavior of AA2219 fabricated by high shear solid state material deposition. Mater Sci Eng A 2018;. Available from: https://doi.org/10.1016/j.msea.2018.03.088.

[14] Rivera OG, Allison PG, Jordon JB, Rodriguez OL, Brewer LN, McClelland Z, et al. Microstructures and mechanical behavior of Inconel 625 fabricated by solid-state additive manufacturing. Mater Sci Eng A 2017;694. Available from: https://doi.org/10.1016/j.msea.2017.03.105.

[15] Avery DZ, Rivera OG, Mason CJT, Phillips BJ, Jordon JB, Su J, et al. Fatigue behavior of solid-state additive manufactured inconel 625. JOM 2018;70(11):2475–84. Available from: https://doi.org/10.1007/s11837-018-3114-7.

[16] Aggen G, Michael Allen C. ASM handbook. Volume I: Properties and selection: Irons, steels, and high-performance alloys, vol. 2. ASM International: The Materials Information Company; 2001. Available from: https://doi.org/10.1016/S0026-0576(03)90166-8.

[17] Theriault A, Xue L, Dryden JR. Fatigue behavior of laser consolidated IN-625 at room and elevated temperatures. Mater Sci Eng A 2009;516:217–25. Available from: https://doi.org/10.1016/j.msea.2009.03.056.

A.1 STRESS-LIFE DATA

A.1.1 2xxx Aluminum Alloys

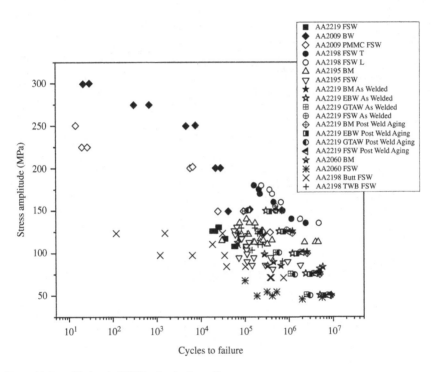

Figure A.1 Stress-life data for 2XXX series aluminum alloy.

Table A.1 Description of Data in Fig. A.1

Legend	Description	Ref.
■ AA2219 FSW	6 mm thick AA2024-T3, 800 rpm, 180 mm/min, $R = 0.1$ at 10 Hz	[1]
◆ AA2009 BM	3 mm thick AA2009-T351 sheet, $R = 0$ at 50 Hz	[2]
◇ 2009 PMMC FSW	3 mm thick AA2009-T351 sheet, 1000 rpm, 50 mm/min, 7 μm SiCp particles, $R = 0$ at 50 Hz	[2]
● AA2198 FSW-T	5 mm thick AA2198-T851, 1000 rpm, 80 mm/min, tool angle of 2 degree, $R = 0.33$ at 80 Hz, transverse to rolling direction	[3]
○ AA2198 FSW-L	5 mm thick AA2198-T851, 1000 rpm, 80 mm/min, tool angle of 2 degree, $R = 0.33$ at 80 Hz, parallel to rolling direction	[3]
△ AA2195 BM	5 mm thick AA2195-T8, $R = 0.1$	[4]
▽ AA2195 FSW	5 mm thick AA2195-T8, 800 rpm, 0.9 mm/s, $R = 0.1$	[4]
★ AA2219 BM As Welded	5 mm thick AA2219-T87, $R = 0$	[5]
☆ AA2219 EBW As Welded	5 mm thick AA2219-T87, electron beam welded with 51 mA direct current, 50 kV, 16 mm/s, 10^{-4} bar vacuum, $R = 0$	[5]
⊞ AA2219 GTAW As Welded	5 mm thick AA2219-T87, gas tungsten arc welded with 150 A alternating current, 30 V, 3 mm/s, shielding gas: 10 L/min pure Argon, $R = 0$	[5]
⊕ AA2219 FSW As Welded	5 mm thick AA2219-T87, 1400 rpm, 1.5 mm/s, $R = 0$	[5]
⊕ AA2219 BM Postweld Aging	5 mm thick AA2219-T87, $R = 0$, postweld aging at 175°C for 12 h	[5]
◨ AA2219 EBW Postweld Aging	5 mm thick AA2219-T87, electron beam welded with 51 mA direct current, 50 kV, 16 mm/s, 10^{-4} bar vacuum, $R = 0$, postweld aging at 175°C for 12 h	[5]
◐ AA2219 GTAW Postweld Aging	5 mm thick AA2219-T87, gas tungsten arc welded with 150 A alternating current, 30 V, 3 mm/s, shielding gas: 10 L/min pure argon, $R = 0$, postweld aging at 175°C for 12 h	[5]
◀ AA2219 FSW Postweld Aging	5 mm thick AA2219-T87, 1400 rpm, 1.5 mm/s, $R = 0$, postweld aging at 175°C for 12 h	[5]
★ AA2060 BM	2 mm thick AA2060-T8, $R = 0.1$ at 30 Hz	[6]
✳ AA2060 FSW	2 mm thick AA2060-T8, 2400 rpm, 100 mm/min, water-cooled FSW, $R = 0.1$ at 30 Hz	[6]
✕ AA2198 Butt FSW	3 mm thick AA2198-T851 600 rpm, 5 mm/s, $R = 0.1$ at 10 Hz	[7]
± AA2198 TWB FSW	3.5–4.2 mm thick AA2198-T851 tailor welded blanks milled to 3 mm, 600 rpm, 5 mm/s, $R = 0.1$ at 10 Hz	[7]

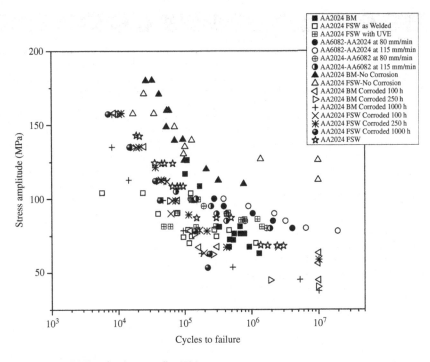

Figure A.2 Stress-life data for aluminum alloy 2024.

Table A.2 Description of Data in Fig. A.2		
Legend	**Description**	**Ref.**
■ AA2024 BM	6 mm thick AA2024-T3, $R = 0.1$ at 87–92 Hz	[8]
□ AA2024 FSW as Welded	6 mm thick AA2024-T3, 800 rpm, 75 mm/min, $R = 0.1$ at 87–92 Hz	[8]
⊞ AA2024 FSW with UVE	6 mm thick AA2024-T3, 800 rpm, 75 mm/min, ultrasonic vibration at 20 kHz with displacement 40 µm applied to surface during welding, $R = 0.1$ at 87–92 Hz	[8]
● AA6061-AA2024 at 80 mm/min	4 mm thick AA6082-T6 and AA2024-T3, 1600 rpm, 80 mm/min, 250 Hz, AA6082 advancing side	[9]
o AA6061-AA2024 at 115 mm/min	4 mm thick AA6082-T6 and AA2024-T3, 1600 rpm, 115 mm/min, 250 Hz, AA6082 advancing side	[9]
⊕ AA2024-AA6082 at 80 mm/min	4 mm thick AA6082-T6 and AA2024-T3, 1600 rpm, 80 mm/min, 250 Hz, AA2024 advancing side	[9]
◑ AA2024-AA6082 at 115 mm/min	4 mm thick AA6082-T6 and AA2024-T3, 1600 rpm, 115 mm/min, 250 Hz, AA2024 advancing side	[9]
▲ AA2024 BM—No Corrosion	4 mm thick AA2024-T3, $R = 0.1$ at 60 Hz	[10]
△ AA2024 FSW—No Corrosion	4 mm thick AA2024-T3, 850 rpm, 300 mm/min, tool angle 0 degree, $R = 0.1$ at 60 Hz	[10]
◁ AA2024 BM—Corroded 100 h	4 mm thick AA2024-T3, $R = 0.1$ at 60 Hz, pre-corroded with 3.5% NaCl for 100 h	[10]

(Continued)

Table A.2 (Continued)		
Legend	Description	Ref.
▷ AA2024 BM—Corroded 250 h	4 mm thick AA2024-T3, $R = 0.1$ at 60 Hz, pre-corroded with 3.5% NaCl for 250 h	[10]
+ AA2024 BM—Corroded 1000 h	4 mm thick AA2024-T3, $R = 0.1$ at 60 Hz, pre-corroded with 3.5% NaCl for 1000 h	[10]
× AA2024 FSW—Corroded 100 h	4 mm thick AA2024-T3, 850 rpm, 300 mm/min, tool angle 0 degree, $R = 0.1$ at 60 Hz, pre-corroded with 3.5% NaCl for 100 h	[10]
✳ AA2024 FSW—Corroded 250 h	4 mm thick AA2024-T3, 850 rpm, 300 mm/min, tool angle 0 degree, $R = 0.1$ at 60 Hz, pre-corroded with 3.5% NaCl for 250 h	[10]
◑ AA2024 FSW—Corroded 1000 h	4 mm thick AA2024-T3, 850 rpm, 300 mm/min, tool angle 0 degree, $R = 0.1$ at 60 Hz, pre-corroded with 3.5% NaCl for 1000 h	[10]
☆ AA2024 FSW	1.6 mm thick AA2024-T3, 700 mm/min, $R = 0.1$	[11]

A.1.2 5xxx Aluminum Alloys

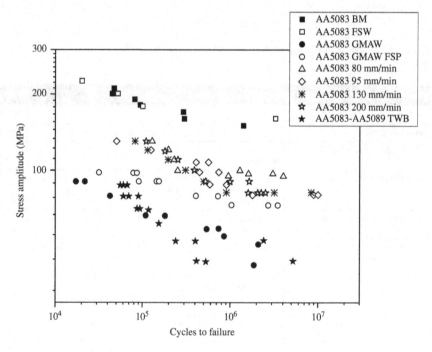

Figure A.3 Stress-life data for aluminum alloy 5083.

Table A.3 Description of Data in Fig. A.3

Data Set	Description	Ref.
■ AA5083 BM	5 mm thick AA5083-O rolled sheet	[12]
□ AA5083 FSW	5 mm thick AA5083-O rolled sheet with friction stir welds	[12]
● AA5083 GMAW	6 mm thick, AA5083-H111 plate welded under gas metal arc welding	[13]
o AA5083 GMAW FSP	6 mm thick, AA5083-H111 plate welded under gas metal arc welding with friction stir processing after the GMAW	[13]
△ AA5083 80 mm/min	8 mm thick As-rolled AA5083-H321 plate friction stir welded with 80 mm/min traverse speed	[14]
◇AA5083 95 mm/min	8 mm thick As-rolled AA5083-H321 plate friction stir welded with 95 mm/min traverse speed	[14]
✳ AA5083 130 mm/min	8 mm thick As-rolled AA5083-H321 plate friction stir welded with 130 mm/min traverse speed	[14]
☆ AA5083 200 mm/min	8 mm thick As-rolled AA5083-H321 plate friction stir welded with 200 mm/min traverse speed	[14]
★ AA5083-AA5089 TWB	8 mm AA5083 friction stir butt welded to 6 mm AA5059 tailor welded blanks	[15]

A.1.3 6xxx Aluminum Alloys

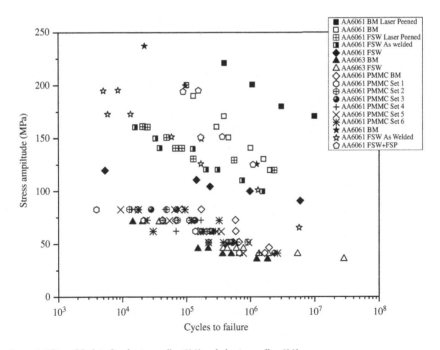

Figure A.4 Stress-life data for aluminum alloy 6061 and aluminum alloy 6063.

Table A.4 Description of Data in Fig. A.4

Legend	Description	Ref.
■ AA6061 BM Laser Peened	3 mm thick plate, $R = -1$ at 22 Hz	[16]
□ AA6061 BM	3 mm thick plate, $R = -1$ at 22 Hz	[16]
⊞ AA6061 FSW Laser Peened	3 mm thick plate, 1400 rpm, 41 mm/min, 3 degree pitch angle, $R = -1$ at 22 Hz	[16]
▨ AA6061 FSW as welded	3 mm thick plate, 1400 rpm, 41 mm/min, 3 degree pitch angle, $R = -1$ at 22 Hz	[16]
◆ AA6061 FSW	5 mm thick plate, 1200 rpm, 200 mm/min, $R = -1$ at 10 Hz	[12]
▲ AA6063 BM	3 mm thick plate, $R = 0.1$ at 8 Hz	[17]
△ AA6063 FSW	3 mm thick plate, 1000 rpm, 9.17 mm/s, $R = 0.1$ at 8 Hz	[17]
◇ AA6061 PMMC BM	4 mm thick plate, $R = 0.1$ at 22 Hz	[18]
○ AA6061 PMMC Set 1	4 mm thick plate, 630 rpm, 115 mm/min, $R = 0.1$ at 22 Hz	[18]
⊕ AA6061 PMMC Set 2	4 mm thick plate, 630 rpm, 170 mm/min, $R = 0.1$ at 22 Hz	[18]
◉ AA6061 PMMC Set 3	4 mm thick plate, 630 rpm, 260 mm/min, $R = 0.1$ at 22 Hz	[18]
+ AA6061 PMMC Set 4	4 mm thick plate, 880 rpm, 115 mm/min, $R = 0.1$ at 22 Hz	[18]
× AA6061 PMMC Set 5	4 mm thick plate, 880 rpm, 170 mm/min, $R = 0.1$ at 22 Hz	[18]
✳ AA6061 PMMC Set 6	4 mm thick plate, 880 rpm, 260 mm/min, $R = 0.1$ at 22 Hz	[18]
★ AA6061 BM	4 mm thick plate, $R = -1$ at 24 Hz	[19]
☆ AA6061 FSW as welded	4 mm thick plate, 1200 rpm, 50 mm/min, pitch angle of 2 degree, $R = -1$ at 24 Hz	[19]
⌂ AA6061 FSW + FSP	4 mm thick plate, 1200 rpm, 50 mm/min, pitch angle of 2 degree, 0.18 mm, $R = -1$ at 24 Hz	[19]

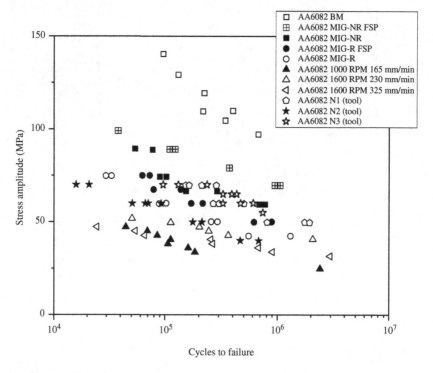

Figure A.5 Stress-life data for aluminum alloy 6082.

Table A.5 Description of Data in Fig. A.5

Legend	Description	Ref.
□ AA6082BM	6 mm thick plate, $R = 0$ at 20–30 Hz	[20]
⊞ AA6082 MIG-NR FSP	6 mm thick plate, 1500 rpm, 240 mm/min, tilt angle of 2.5 degree, $R = 0$ at 20–30 Hz	[20]
■ AA6082 MIG-NR	6 mm thick plate, $R = 0$ at 20–30 Hz	[20]
● AA6082 MIG-R FSP	6 mm thick plate, 1500 rpm, 240 mm/min, tilt angle of 2.5 degree, $R = 0$ at 20–30 Hz	[20]
○ AA6082 MIG-R	6 mm thick plate, $R = 0$ at 20–30 Hz	[20]
▲ AA6082 1000 rpm 165 mm/min	4 mm thick plate, tilt angle 3 degree, $R = 0.1$ at up to 100 Hz	[21]
△ AA6082 1600 rpm 230 mm/min	4 mm thick plate, tilt angle 3 degree, $R = 0.1$ at up to 100 Hz	[21]
◁ AA6082 1600 rpm 325 mm/min	4 mm thick plate, tilt angle 3 degree, $R = 0.1$ at up to 100 Hz	[21]
⬠ AA6082 N1 (tool)	8 mm thick plate, machined weld face surface, 710 rpm, 900 mm/min, tilt angle 1.5 degree, $R = 0.2$ at 20 Hz	[22]
★ AA6082 N2 (tool)	8 mm thick plate, machined weld face surface, 710 rpm, 900 mm/min, tilt angle 1.5 degree, $R = 0.2$ at 20 Hz	[22]
☆ AA6082 N3 (tool)	8 mm thick plate, machined weld face surface, 710 rpm, 900 mm/min, tilt angle 1.5 degree, $R = 0.2$ at 20 Hz	[22]

A.1.4 7xxx Aluminum Alloys

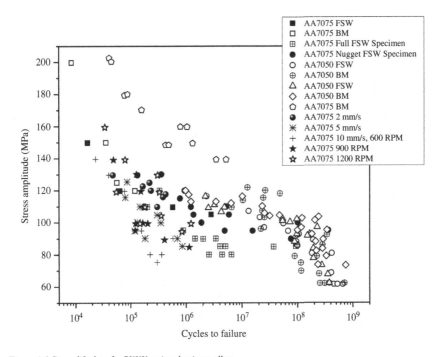

Figure A.6 Stress-life data for 7XXX series aluminum alloy.

Table A.6 Description of Data in Fig. A.6

Legend	Description	Ref.
■ AA7075 FSW	5 mm thick plate, 1000 rpm, 350 mm/min, $R = -1$ at 10 Hz	[12]
□ AA7075 BM	5 mm thick plate, $R = -1$ at 10 Hz	[12]
⊞ AA7075 Full FSW Specimen	1 mm thick plate, 300 rpm, 150 mm/min, 8 degree tilt angle, $R = -1$ at 20 Hz	[23]
● AA7075 Nugget FSW Specimen	1 mm thick plate, 300 rpm, 150 mm/min, 8 degree tilt angle, $R = -1$ at 20 Hz	[23]
○ AA7050 FSW	12 mm thick plate, 300 rpm, 95 mm/min, $R = -1$ at 20 kHz	[24]
⊕ AA7050 BM	12 mm thick plate, $R = -1$ at 20 kHz	[24]
△ AA7050 FSW	12 mm thick plate, 800 rpm, 2.5 mm/s, 2.5 degree tilt angle, $R = -1$ at 19 kHz	[25]
◇ AA7050 BM	12 mm thick plate, $R = -1$ at 19 kHz	[25]
◌ AA7075 BM	2.3 mm thick plate, $R = 0.1$ at 20 Hz	[26]
◐ AA7075 2 mm/s	2.3 mm thick plate, 82 degree angle from axis of tool, 600 rpm, $R = 0.1$ at 20 Hz	[26]
✳ AA7075 5 mm/s	2.3 mm thick plate, 82 degree angle from axis of tool, 600 rpm, $R = 0.1$ at 20 Hz	[26]
+ AA7075 10 mm/s, 600 rpm	2.3 mm thick plate, 82 degree angle from axis of tool, $R = 0.1$ at 20 Hz	[26]
★ AA7075 900 rpm	2.3 mm thick plate, 82 degree angle from axis of tool, 10 mm/s, $R = 0.1$ at 20 Hz	[26]
☆ AA7075 1200 rpm	2.3 mm thick plate, 82 degree angle from axis of tool, 10 mm/s, $R = 0.1$ at 20 Hz	[26]

A.1.5 Miscellaneous Aluminum Alloys

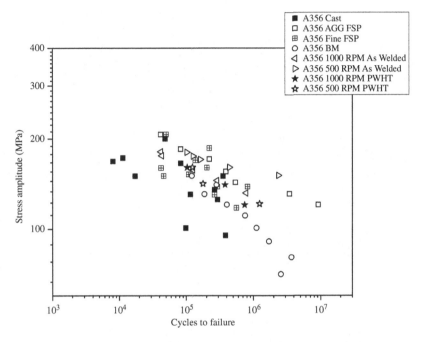

Figure A.7 Stress-life data for aluminum alloy A356.

Table A.7 Description of Data in Fig. A.7		
Data Set	**Description**	**Ref.**
■ A356 Cast	A356 cast material heat treated to the T6 condition	[27]
□ A356 AGG FSP	A356 friction stir welded with a rotation rate of 1500 rpm and 102 mm/min	[27]
⊞ A356 Fine FSP	A356 friction stir welded with a rotation rate of 300 rpm and 102 mm/min	[27]
o A356 BM	190 × 33 × 39 mm cast ingots A356-T6	[28]
◁ A356 1000 rpm As Welded	190 × 33 × 39 mm cast ingots A356-T6 friction stir welded at 1000 rpm	[28]
▷ A356 500 rpm As Welded	190 × 33 × 39 mm cast ingots A356-T6 friction stir welded at 500 rpm	[28]
★ A356 1000 rpm PWHT	190 × 33 × 39 mm cast ingots A356-T6 friction stir welded at 1000 rpm with a postweld heat treatment	[28]
☆ A356 500 rpm PWHT	190 × 33 × 39 mm cast ingots A356-T6 friction stir welded at 500 rpm with a postweld heat treatment	[28]

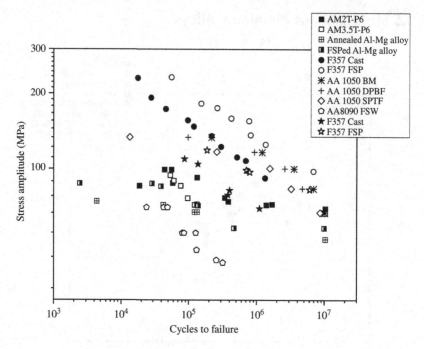

Figure A.8 Stress-life data for miscellaneous aluminum alloys.

Table A.8 Description of Data in Fig. A.8		
Data Set	**Description**	**Ref.**
■ AM2T-P6	AA5052 with 2 vol% TiO_2 nanoparticles	[29]
□ AM3.5T-P6	AA5052 with 3.5 vol% TiO_2 nanoparticles	[29]
⊞ Annealed Al-Mg alloy	Annealed AA5052 sheet	[29]
◧ FSPed Al-Mg alloy	Annealed AA5052 sheet that has been friction stir processed	[29]
● F357 Cast	3.3 mm thick investment cast F357 plates heat treated to the T6 condition	[30]
○ F357 FSP	3.3 mm thick investment cast F357 plates friction stir welded heat treated to the T6 condition	[30]
✳ AA1050 BM	AA1050 of various plate thicknesses were used in this study	[31]
+ AA1050 DPBF	Double pass bottom surface flat friction stir welded AA1050 plate samples	[31]
◇ AA1050 SPTF	Single pass top surface flat friction stir welded AA1050 plate samples	[31]
⬠ AA8090 FSW	AA8090-T8 plates friction stir welded with various rotation speed and translation speed	[32]
★ F357 Cast	3.3 mm thick investment cast F357	[33]
☆ F357 FSP	Friction stir processed 3.3 mm thick investment cast F357	[33]

A.1.6 Magnesium Alloys

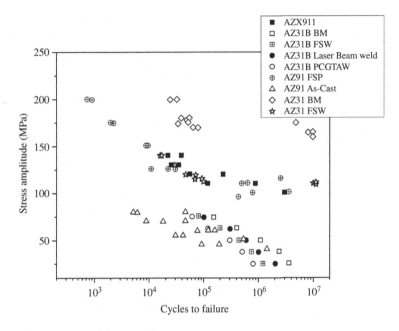

Figure A.9 Stress-life data for magnesium alloys.

Table A.9 Description of Data in Fig. A.9

Data Sets	Descriptions	Ref.
■ AZX911	4 mm thick AZX911 hot extruded sheet butt welded along the extrusion direction with friction stir welding parameters of 1000 rpm, 200 mm/min, and 3 degree tilt angle	[34]
□ AZ31B BM	6 mm thick AZ31B rolled plate	[35]
⊞ AZ31B FSW	6 mm thick AZ31B rolled plate butt friction stir welded with parameters 40 mm/min and 1600 rpm	[35]
● AZ31B Laser Beam Weld	6 mm thick AZ31B rolled plate butt laser beam welded with parameters 5500 mm/min, 2500 W, −1.5 mm, and 0.027 kJ/mm	[35]
○ AZ31B PCGTAW	6 mm thick AZ31B rolled plate butt pulsed current gas tungsten arc welding with parameters 180 mm/min, 210 A peak current, 80 A base current, 6 Hz pulse frequency, and 0.85 kJ/mm	[35]
⊕ AZ91 FSP	8 mm thick AZ91D plate machined from cast billets friction stir processed with two passes with a rotation rate of 400 rpm, traversing speed of 100 mm/min	[36]
△ AZ91 As-Cast	8 mm thick AZ91D plate machined from cast billets	[36]
◇ AZ31 BM	6.4 mm thick AZ31-H24 plate	[37]
☆ AZ31 FSW	6.4 mm thick AZ31-H24 friction stir welded at 100 mm/min, rotation rate of 800 rpm, and 2.8 degree tilt angle	[37]

A.1.7 Dissimilar Alloys

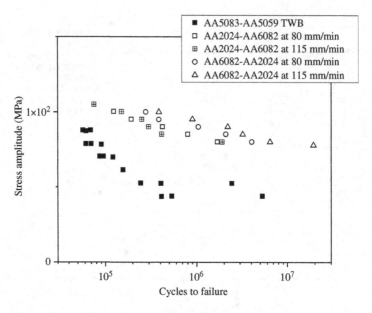

Figure A.10 Stress-life data for dissimilar metals.

Table A.10 Description of Data in Fig. A.10		
Data Sets	**Descriptions**	**Ref.**
■ AA5083-5059 TWB	8 mm AA5083 friction stir butt welded to 6 mm AA5059 tailor welded blanks	[15]
□ AA2024-6082 at 80 mm/min	4 mm thick AA2024-T3 (advancing side) friction stir butt welded to 4 mm thick AA6082-T6 (retreating side) with a traverse speed of 80 mm/min	[9]
⊞ AA2024-6082 at 115 mm/min	4 mm thick AA2024-T3 (advancing side) friction stir butt welded to 4 mm thick AA6082-T6 (retreating side) with a traverse speed of 115 mm/min	[9]
o AA6082-2024 at 80 mm/min	4 mm thick AA6082-T6 (advancing side) friction stir butt welded to 4 mm thick AA2024-T3 (retreating side) with a traverse speed of 80 mm/min	[9]
△ AA6082-2024 at 115 mm/min	4 mm thick AA6082-T6 (advancing side) friction stir butt welded to 4 mm thick AA2024-T3 (retreating side) with a traverse speed of 115 mm/min	[9]

A.1.8 Steels

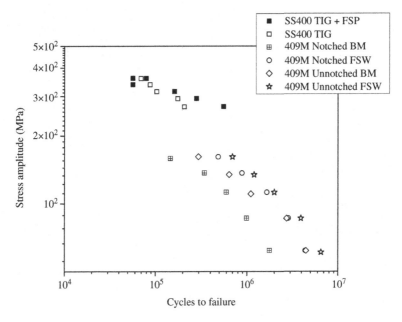

Figure A.11 Stress-life data for steels.

Table A.11 Description of Data in Fig. A.11		
Data Set	**Description**	**Ref.**
■ SS400 TIG + FSP	5 mm thick SS400 plate TIG welded then friction stir processed over the TIG weld bead	[38]
□ SS400 TIG	5 mm thick SS400 plate TIG welded	[38]
⊞ 409 M Notched BM	4 mm thick, cold rolled, annealed, and pickled AISI 409 M ferritic stainless steel plates in the notched specimen geometry	[39]
o 409 M Notched FSW	4 mm thick, cold rolled, annealed, and pickled AISI 409 M ferritic stainless steel plates in the notched specimen geometry with a friction stir weld running through the gage section	[39]
◇ 409 M Unnotched BM	4 mm thick, cold rolled, annealed, and pickled AISI 409 M ferritic stainless steel plates in the unnotched specimen geometry	[39]
☆ 409 M Unnotched FSW	4 mm thick, cold rolled, annealed, and pickled AISI 409 M ferritic stainless steel plates in the unnotched specimen geometry with a friction stir weld running through the gage section	[39]

A.1.9 Titanium Alloys

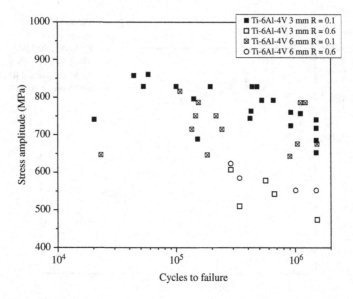

Figure A.12 Stress-life data for Ti-6Al-4V.

Table A.12 Description of Data in Fig. A.12		
Data Set	**Description**	**Ref.**
■ Ti-6Al-4V 3 mm $R = 0.1$	3 mm thick Ti-6Al-4V sheet tested at a load ratio of $R = 0.1$	[40]
□ Ti-6Al-4V 3 mm $R = 0.6$	3 mm thick Ti-6Al-4V sheet tested at a load ratio of $R = 0.6$	[40]
⊠ Ti-6Al-4V 6 mm $R = 0.1$	6 mm thick Ti-6Al-4V plate tested at a load ratio of $R = 0.1$	[40]
o Ti-6Al-4V 6 mm $R = 0.6$	6 mm thick Ti-6Al-4V plate tested at a load ratio of $R = 0.6$	[40]

A.2 STRAIN-LIFE DATA

A.2.1 2xxx Aluminum Alloys

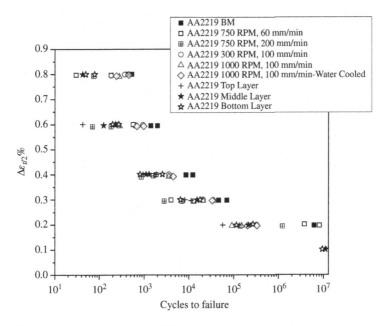

Figure A.13 Strain-life data for aluminum alloy 2219.

Table A.13 Description of Data in Fig. A.13		
Legend	**Description**	**Ref.**
■ AA2219 BM	6 mm thick plate, 2.8 degree tilt angle, $R = -1$ at 50 Hz	[41]
□ AA2219 750 rpm, 60 mm/min	6 mm thick plate, $R = -1$ at 50 Hz	[41]
⊞ AA2219 750 rpm, 200 mm/min	6 mm thick plate, 2.8 degree tilt angle, $R = -1$ at 50 Hz	[41]
o AA2219 300 rpm, 100 mm/min	6 mm thick plate, 2.8 degree tilt angle, $R = -1$ at 50 Hz	[41]
△ AA2219 1000 rpm, 100 mm/min	6 mm thick plate, 2.8 degree tilt angle, $R = -1$ at 50 Hz	[41]
◇ AA2219 1000 rpm, 100 mm/min— Water Cooled	6 mm thick plate, 2.8 degree tilt angle, $R = -1$ at 50 Hz	[41]
+ AA2219 Top Layer	20 mm thick plate, 300 rpm, 100 mm/min, 2.8 degree tilt angle, $R = -1$ at 50 Hz	[42]
★ AA2219 Middle Layer	20 mm thick plate, 300 rpm, 100 mm/min, 2.8 degree tilt angle, $R = -1$ at 50 Hz	[42]
☆ AA2219 Bottom Layer	20 mm thick plate, 300 rpm, 100 mm/min, 2.8 degree tilt angle, $R = -1$ at 50 Hz	[42]

A.2.2 6xxx Aluminum Alloys

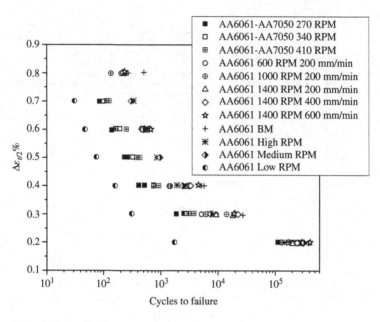

Figure A.14 Strain-life data for aluminum alloy 6061.

Table A.14 Description of Data in Fig. A.14		
Legend	Description	Ref.
■ AA6061-AA7050 270 rpm	5 mm thick plate, 270 rpm, 114 mm/min, $R = -1$ at 5 Hz	[43]
□ AA6061-AA7050 340 rpm	5 mm thick plate, 340 rpm, 114 mm/min, $R = -1$ at 5 Hz	[43]
⊞ AA6061-AA7050 410 rpm	5 mm thick plate, 410 rpm, 114 mm/min, $R = -1$ at 5 Hz	[43]
o AA6061 600 rpm 200 mm/min	6.2 mm thick plate, 600 rpm, 200 mm/min, $R = -1$ at 50 Hz	[44]
⊕ AA6061 1000 rpm 200 mm/min	6.2 mm thick plate, 1000 rpm, 200 mm/min, $R = -1$ at 50 Hz	[44]
△ AA6061 1400 rpm 200 mm/min	6.2 mm thick plate, 1400 rpm, 200 mm/min, $R = -1$ at 50 Hz	[44]
◇ AA6061 1400 rpm 400 mm/min	6.2 mm thick plate, 1400 rpm, 400 mm/min, $R = -1$ at 50 Hz	[44]
☆ AA6061 1400 rpm 600 mm/min	6.2 mm thick plate, 1400 rpm, 600 mm/min, $R = -1$ at 50 Hz	[44]
+ AA6061 BM	6.2 mm thick plate, $R = -1$ at 50 Hz	[44]
✳ AA6061 High rpm	5.6 mm thick rolled plate, 460 rpm, 254 mm/min, $R = -1$	[45]
◆ AA6061 Medium rpm	5.6 mm thick rolled plate, 380 rpm, 254 mm/min, $R = -1$	[45]
◑ AA6061 Low rpm	5.6 mm thick rolled plate, 300 rpm, 254 mm/min, $R = -1$	[45]

A.2.3 7xxx Aluminum Alloys

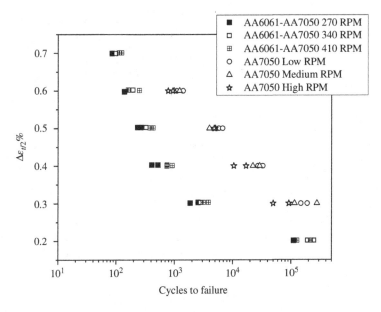

Figure A.15 Strain-life data for 7XXX series aluminum alloy.

Table A.15 Description of Data in Fig. A.15

Legend	Description	Ref.
■ AA6061-AA7050 270 rpm	5 mm thick plate, 270 rpm, 114 mm/min, $R = -1$ at 5 Hz	[43]
□ AA6061-AA7050 340 rpm	5 mm thick plate, 340 rpm, 114 mm/min, $R = -1$ at 5 Hz	[43]
⊞ AA6061-AA7050 410 rpm	5 mm thick plate, 410 rpm, 114 mm/min, $R = -1$ at 5 Hz	[43]
o AA7050 Low rpm	5 mm thick plate, 240 rpm, 152 mm/min, immersed in 3.5% NaCl solution for 15 and 30 days	[46]
△ AA7050 Medium rpm	5 mm thick plate, 300 rpm, 152 mm/min, immersed in 3.5% NaCl solution for 15 and 30 days	[46]
☆ AA7050 High rpm	5 mm thick plate, 360 rpm, 152 mm/min, immersed in 3.5% NaCl solution for 15 and 30 days	[46]

A.2.4 Magnesium Alloys

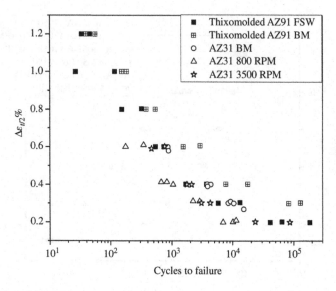

Figure A.16 Strain-life data for magnesium alloys.

Table A.16 Description of Data in Fig. A.16		
Legend	Description	Ref.
■ Thixomolded AZ91 FSW	3 mm thick plate, 800 rpm, 50 mm/min, 2.7 degree tilt angle, $R = -1$ at 50 Hz	[47]
⊞ Thixomolded AZ91 BM	3 mm thick plate, $R = -1$ at 50 Hz	[47]
○ AZ31 BM	6.4 mm thick rolled plate	[48]
△ AZ31 800 rpm	6.4 mm thick rolled plate, 100 mm/min, 800 rpm	[48]
☆ AZ31 3500 rpm	6.4 mm thick rolled plate, 100 mm/min, 3500 rpm	[48]

A.3 FATIGUE CRACK GROWTH DATA

A.3.1 2xxx Aluminum Alloys

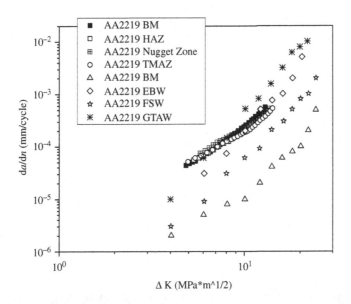

Figure A.17 Crack growth data for aluminum alloy 2219.

Table A.17 Description of Data in Fig. A.17		
Data Sets	Description	Ref.
■ AA2219 BM	Aluminum plate AA2219-T6	[49]
□ AA2219 HAZ	Aluminum plate AA2219-T6 friction stir welded at 800 rpm and 150 mm/min with the crack path in the HAZ	[49]
⊞ AA2219 Nugget Zone	Aluminum plate AA2219-T6 friction stir welded at 800 rpm and 150 mm/min with the crack path in the nugget zone	[49]
o AA2219 TMAZ	Aluminum plate AA2219-T6 friction stir welded at 800 rpm and 150 mm/min with the crack path in the TMAZ	[49]
△ AA2219 BM	Rolled AA2219-T87 plate machined to a thickness of 5 mm	[50]
◇ AA2219 EBW	Rolled AA2219-T87 plate machined to a thickness of 5 mm electron beam butt welded	[50]
☆ AA2219 FSW	Rolled AA2219-T87 plate machined to a thickness of 5 mm friction stir butt welded	[50]
✳ AA2219 GTAW	Rolled AA2219-T87 plate machined to a thickness of 5 mm gas tungsten arc butt welded	[50]

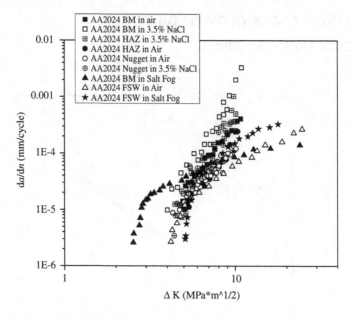

Figure A.18 Crack growth data for aluminum alloy 2024.

Table A.18 Description of Data in Fig. A.18		
Data Sets	**Description**	**Ref.**
■ AA2024 BM in air	Hot-rolled AA2024 subjected to equal channel angular pressing in an air environment	[51]
□ AA2024 BM in 3.5% NaCl	Hot-rolled AA2024 subjected to equal channel angular pressing in a 3.5 wt.% NaCl air solution	[51]
⊞ AA2024 HAZ in 3.5% NaCl	Hot-rolled AA2024 subjected to equal channel angular pressing friction stir welded with crack paths through the HAZ in a 3.5 wt.% NaCl air solution	[51]
● AA2024 HAZ in Air	Hot-rolled AA2024 subjected to equal channel angular pressing friction stir welded with crack paths through the HAZ in air environment	[51]
○ AA2024 Nugget in Air	Hot-rolled AA2024 subjected to equal channel angular pressing friction stir welded with crack paths through the weld nugget in air environment	[51]
⊕ AA2024 Nugget in 3.5% NaCl	Hot-rolled AA2024 subjected to equal channel angular pressing friction stir welded with crack paths through the weld nugget in a 3.5 wt.% NaCl air solution	[51]
▲ AA2024 BM in Salt Fog	3.2 mm thick AA2024-T3 sheet tested in a 3.5 wt.% NaCl air solution	[52]
△ AA2024 FSW in Air	3.2 mm thick AA2024-T3 sheet friction stir welded tested in air environment	[52]
★ AA2024 FSW in Salt Fog	3.2 mm thick AA2024-T3 sheet friction stir welded tested in a 3.5 wt.% NaCl air solution	[52]

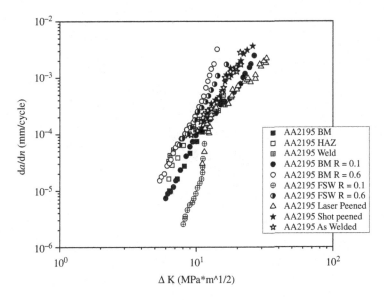

Figure A.19 Crack growth data for aluminum alloy 2195.

Table A.19 Description of Data in Fig. A.19

Data Sets	Descriptions	Ref.
■ AA2195 BM	AA2195-T8X plates machined to 5 mm	[53]
□ AA2195 HAZ	AA2195-T8X plates machined to 5 mm with the HAZ of the friction stir weld along the crack path	[53]
⊞ AA2195 Weld	AA2195-T8X plates machined to 5 mm with the friction stir weld along the crack path	[53]
● AA2195 BM $R = 0.1$	Rolled AA2195-T8 plate machined to 8 mm tested at a load ratio of $R = 0.1$	[54]
○ AA2195 BM $R = 0.6$	Rolled AA2195-T8 plate machined to 8 mm tested at a load ratio of $R = 0.6$	[54]
⊕ AA2195 FSW $R = 0.1$	Rolled AA2195-T8 plate machined to 8 mm friction stir welded in the direction of the to the rolling direction tested at a load ratio of $R = 0.1$	[54]
◐ AA2195 FSW $R = 0.6$	Rolled AA2195-T8 plate machined to 8 mm friction stir welded in the direction of the to the rolling direction tested at a load ratio of $R = 0.6$	[54]
△ AA2195 Laser Peened	12.5 mm thick AA2195-T8 plates friction stir welded at 300 rpm and 150 mm/min laser peened after welding	[55]
★ AA2195 Shot Peened	12.5 mm thick AA2195-T8 plates friction stir welded at 300 rpm and 150 mm/min shot peened after welding	[55]
☆ AA2195 As Welded	12.5 mm thick AA2195-T8 plates friction stir welded at 300 rpm and 150 mm/min with no postprocessing	[55]

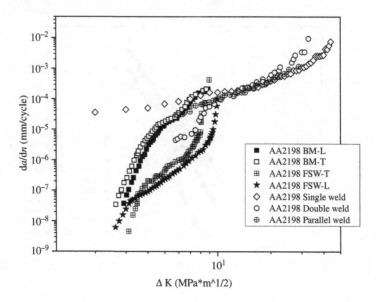

Figure A.20 Crack growth data for aluminum alloy 2198.

Table A.20 Description of Data in Fig. A.20		
Data Sets	**Descriptions**	**Ref.**
■ AA2198 BM-L	5 mm thick rolled AA2198-T851 sheet in the longitudinal direction	[56]
□ AA2198 BM-T	5 mm thick rolled AA2198-T851 sheet in the transverse direction	[56]
⊞ AA2198 FSW-T	5 mm thick rolled AA2198-T851 sheet in the transverse direction friction stir welded with a rotation speed of 1000 rpm and 80 mm/min	[56]
★ AA2198 FSW-L	5 mm thick rolled AA2198-T851 sheet in the longitudinal direction friction stir welded with a rotation speed of 1000 rpm and 80 mm/min	[56]
◇ AA2198 Single Weld	1.6 and 3.2 mm thick AA2198-T8 sheets friction stir welded perpendicular to the initial crack	[57]
○ AA2198 Double Weld	1.6 and 3.2 mm thick AA2198-T8 sheets friction stir welded twice perpendicular along either side of the initial crack	[57]
⊕ AA2198 Parallel Weld	1.6 and 3.2 mm thick AA2198-T8 sheets friction stir welded parallel to the initial crack	[57]

A.3.2 5xxx Aluminum Alloys

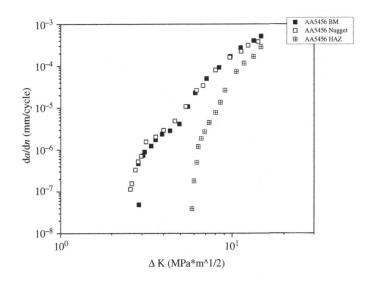

Figure A.21 Crack growth data for aluminum alloy 5456.

Table A.21 Description of Data in Fig. A.21		
Data Sets	**Description**	**Ref.**
■ AA5456 BM	12.7 mm thick AA5456-H116 plate	[58]
□ AA5456 Nugget	12.7 mm thick AA5456-H116 plate friction stir butt welded at 480 rpm and 5 mm/s with the crack path running through the center of the weld	[58]
⊞ AA5456 HAZ	12.7 mm thick AA5456-H116 plate friction stir butt welded at 480 rpm and 5 mm/s with the crack path running heat affected zone	[58]

A.3.3 6xxx Aluminum Alloys

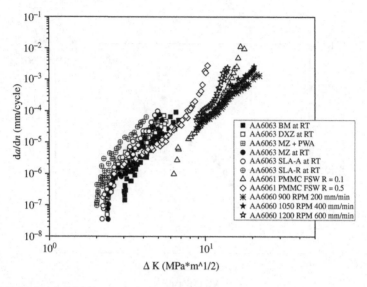

Figure A.22 Crack growth data for aluminum alloy 6xxx.

Table A.22 Description of Data in Fig. A.22		
Data Set	**Description**	**Ref.**
■ AA6063 BM at RT	AA6063-T5 base material	[59]
□ AA6063 DXZ at RT	AA6063-T5 friction stir welded with the machined notch in the weld nugget	[59]
⊞ AA6063 MZ + PWA	AA6063-T5 Friction Stir Welded with the machined notch 5 mm away from the weld nugget with a postweld aging treatment	[59]
● AA6063 MZ at RT	AA6063-T5 friction stir welded with the machined notch 5 mm away from the weld nugget	[59]
○ AA6063 SLA-A at RT	AA6063-T5 friction stir welded with the machined notch 8.5 mm away from the weld nugget on the advancing side	[59]
⊕ AA6063 SLA-R at RT	AA6063-T5 friction stir welded with the machined notch 8.5 mm away from the weld nugget on the retreating side	[59]
△ AA6061 PMMC FSW $R = 0.1$	AA6061 reinforced with 20 vol.% Al_2O_3 friction stir butt welded and tested with a load ratio of $R = 0.1$	[60]
◇ AA6061 PMMC FSW $R = 0.5$	AA6061 reinforced with 20 vol.% Al_2O_3 friction stir butt welded and tested with a load ratio of $R = 0.5$	[60]
✳ AA6060 900 rpm 200 mm/min	8 mm thick AA6060-T6 plate friction stir welded at 900 rpm and 200 mm/min	[61]
★ AA6060 1050 rpm 400 mm/min	8 mm thick AA6060-T6 plate friction stir welded at 1050 rpm and 400 mm/min	[61]
☆ AA6060 1200 rpm 600 mm/min	8 mm thick AA6060-T6 plate friction stir welded at 1200 rpm and 600 mm/min	[61]

A.3.4 7xxx Aluminum Alloys

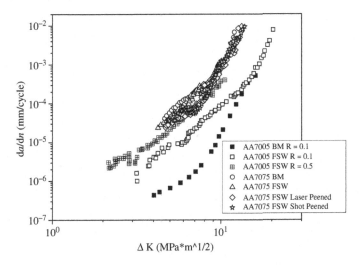

Figure A.23 Crack growth data for 7XXX series aluminum alloy.

Table A.23 Description of Data in Fig. A.23

Data Sets	Description	Ref.
■ AA7005 BM $R = 0.1$	Extruded AA7005-T6 reinforced with 10 vol.% Al_2O_3 particles tested at a load ratio of $R = 0.1$	[62]
□ AA7005 FSW $R = 0.1$	Extruded AA7005-T6 reinforced with 10 vol.% Al_2O_3 particles friction stir welded at 600 rpm and 300 mm/min tested at a load ratio of $R = 0.1$	[62]
⊞ AA7005 FSW $R = 0.5$	Extruded AA7005-T6 reinforced with 10 vol.% Al_2O_3 particles friction stir welded at 600 rpm and 300 mm/min tested at a load ratio of $R = 0.5$	[62]
○ AA7075 BM	12.5 mm AA7075-T7351 thick plate	[63]
△ AA7075 FSW	12.5 mm AA7075-T7351 thick plate friction stir welded at 300 rpm and 150 mm/min	[63]
◇ AA7075 FSW Laser Peened	12.5 mm AA7075-T7351 thick plate friction stir welded at 300 rpm and 150 mm/min laser peened after weld	[63]
☆ AA7075 FSW Shot Peened	12.5 mm AA7075-T7351 thick plate friction stir welded at 300 rpm and 150 mm/min shot peened after weld	[63]

A.3.5 Magnesium Alloys

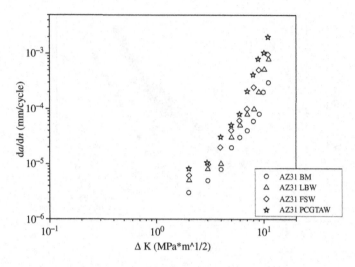

Figure A.24 Crack growth data for magnesium alloy.

Table A.24 Description of Data in Fig. A.24		
Data Sets	**Description**	**Ref.**
○ AZ31 BM	6 mm thick rolled AZ31B plate	[35]
△ AZ31 BM	6 mm thick rolled AZ31B plate laser beam welded at 5500 mm/min, 2500 W, −1.5 mm focal point, and 0.027 kJ/mm heat input	[35]
◇ AZ31 FSW	6 mm thick rolled AZ31B plate friction stir welded at 1600 rpm, 40 mm/min, and 3 kN axial force	[35]
☆ AZ31 PCGTAW	6 mm thick rolled AZ31B plate particularly gas tungsten arc welded at 180 mm/min, 210 A peak current, 80 A base current, 6 Hz pulse frequency, and 0.85 kJ/mm heat input	[35]

A.3.6 Titanium Alloys

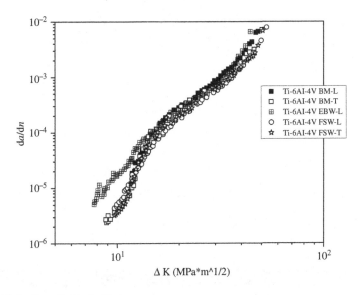

Figure A.25 Crack growth data for titanium.

Table A.25 Description of Data in Fig. A.25

Legend	Description	Ref.
■ Ti-6Al-4V BM-L	24 mm thick Ti-6Al-4V plate in the longitudinal direction	[64]
□ Ti-6Al-4V BM-T	24 mm thick Ti-6Al-4V plate in the transverse direction	[64]
⊞ Ti-6Al-4V EBW-L	24 mm thick Ti-6Al-4V plate in the longitudinal direction electron beam welded with a voltage of 45 kV, 50 MA of beam current, and a travel speed of 750 mm/min	[64]
o Ti-6Al-4V FSW-L	24 mm thick Ti-6Al-4V plate in the longitudinal direction friction stir welded at 150 rpm and 75 mm/min	[64]
☆ Ti-6Al-4V FSW-T	24 mm thick Ti-6Al-4V plate in the transverse direction friction stir welded at 150 rpm and 75 mm/min	[64]

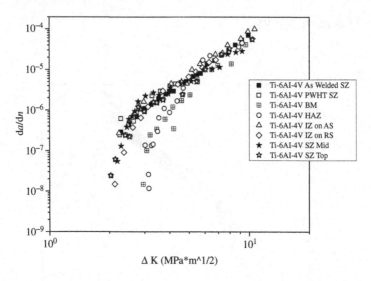

Figure A.26 Crack growth data for titanium based on location.

Table A.26 Description of Data in Fig. A.26		
Data Sets	**Description**	**Ref.**
■ Ti-6Al-4V As Welded SZ	7 mm thick Ti-6Al-4V plate friction stir welded at 120 rpm and 15 mm/min with a crack path in the stir zone	[65]
□ Ti-6Al-4V PWHT SZ	7 mm thick Ti-6Al-4V plate friction stir welded at 120 rpm and 15 mm/min with a crack path in the stir zone with a postweld heat treatment	[65]
⊞ Ti-6Al-4V BM	7 mm thick Ti-6Al-4V plate	[65]
o Ti-6Al-4V HAZ	7 mm thick Ti-6Al-4V plate friction stir welded at 120 rpm and 15 mm/min with a crack path in the heat affected zone	[65]
△ Ti-6Al-4V IZ on As	7 mm thick Ti-6Al-4V plate friction stir welded at 120 rpm and 15 mm/min with a crack path in the interfacial zone of the advancing side	[65]
◇ Ti-6Al-4V IZ on RS	7 mm thick Ti-6Al-4V plate friction stir welded at 120 rpm and 15 mm/min with a crack path in the interfacial zone of the retreating side	[65]
★ Ti-6Al-4V SZ Mid	7 mm thick Ti-6Al-4V plate friction stir welded at 120 rpm and 15 mm/min with a crack path in the stir zone at a depth of 2.4 mm	[65]
☆ Ti-6Al-4V SZ Top	7 mm thick Ti-6Al-4V plate friction stir welded at 120 rpm and 15 mm/min with a crack path in the stir zone at a depth of 0.4 mm	[65]

A.3.7 Steels

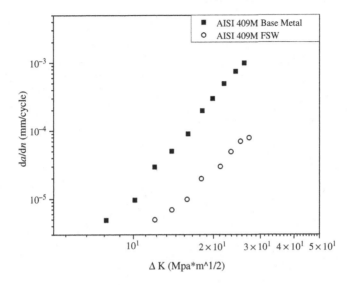

Figure A.27 Crack growth data for steel.

Table A.27 Description of Data in Fig. A.27		
Data Set	Description	Ref.
■ AISI 409 M Base Metal	4 mm thick, cold rolled, annealed, and pickled AISI 409 M stainless steel plate	[39]
○ AISI 409 M FSW	4 mm thick, cold rolled, annealed, and pickled AISI 409 M stainless steel plate Friction Stir Welded at 1000 rpm and 90 mm/min	[39]

A.4 OVERLAP WELDS

A.4.1 FSSW Aluminum Alloys

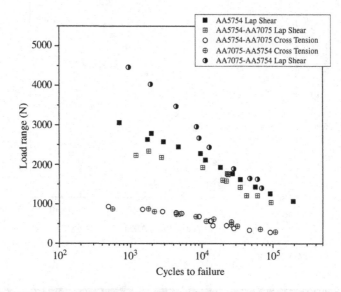

Figure A.28 Fatigue data for FSSW of 5xxx series aluminum alloys.

Table A.28 Description of Data in Fig. A.28		
Legend	Description	Ref.
■ AA5754 Lap Shear	2.0 mm thick plates, 3000 rpm, $R = 0.2$ at 10 Hz	[66]
⊞ AA5754-AA7075 Lap Shear	2.0 mm thick top plate, 1.6 mm thick bottom plate	[67]
o AA5754-AA7075 Cross Tension	2.0 mm thick top plate, 1.6 mm thick bottom plate	[67]
⊕ AA7075-AA5754 Cross Tension	1.6 mm thick top plate, 2.0 mm thick bottom plate	[67]
◑ AA7075-AA5754 Lap Shear	1.6 mm thick top plate, 2.0 mm thick bottom plate	[67]

A.4.2 FSSW Dissimilar Alloys

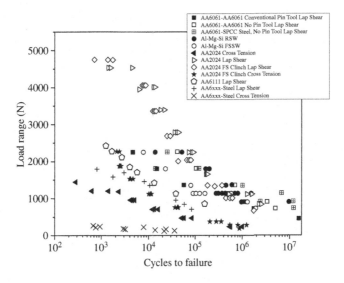

Figure A.29 Fatigue data for FSSW of miscellaneous aluminum alloys.

Table A.29 Description of Data in Fig. A.29

Legend	Description	Ref.
■ AA6061-AA6061 Conventional Pin Tool Lap Shear	2 mm thick plates, 3000 rpm, 10 mm/min, $R = 0.1$ at 10 Hz	[68]
□ AA6061-AA6061 No Pin Tool Lap Shear	2 mm thick plates, 3000 rpm, 10 mm/min, $R = 0.1$ at 10 Hz	[68]
⊞ AA6061-SPCC Steel, No Pin Tool Lap Shear	2 mm thick plates, 3000 rpm, 10 mm/min, $R = 0.1$ at 10 Hz	[68]
● Al-Mg-Si RSW	2 mm thick plates, $R = -0.1$ at 10 Hz	[69]
o Al-Mg-Si FSSW	2 mm thick plates, $R = -0.1$ at 10 Hz	[69]
◄ AA2024 Cross Tension	1.6 mm thick plates, 500 rpm, 120 mm/min indentation rate, 2.8 mm indentation depth, $R = 0.1$ at 5 Hz	[70]
▷ AA2024 Lap Shear	1.6 mm thick plates, 500 rpm, 120 mm/min indentation rate, 2.8 mm indentation depth, $R = 0.1$ at 5 Hz	[70]
◇ AA2024 FS Clinch Lap Shear	1.6 mm thick plates, 1000 rpm, 4.2 mm punching depth, 1.0 mm/s punching rate, 2.5 s dwell time	[71]
★ AA2024 FS Clinch Cross Tension	1.6 mm thick plates, 1000 rpm, 4.2 mm punching depth, 1.0 mm/s punching rate, 2.5 s dwell time	[71]
⟐ AA6111 Lap Shear	0.94 mm thick top plate, 1.04 mm thick bottom plate, 3000 rpm, $R = 0.2$ at 10 Hz	[66]
+ AA6111-Steel Lap Shear	1.3 mm thick 6xxx plate, 0.8 mm thick steel plate, 1500 rpm, 0.85 plunge depth, 5 s dwell time	[72]
× AA6111-Steel Cross Tension	1.3 mm thick 6xxx plate, 0.8 mm thick steel plate, 1500 rpm, 0.85 plunge depth, 5 s dwell time	[72]

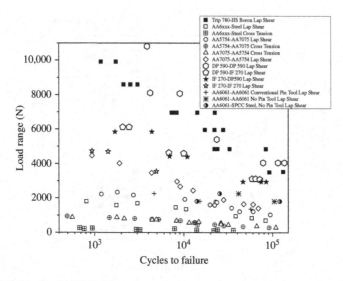

Figure A.30 Fatigue data for FSSW of dissimilar metals.

Table A.30 Description of Data in Fig. A.30

Legend	Description	Ref.
■ Trip 780-HS Boron Lap Shear	1.5 mm thick Trip 780 plate, 1.4 mm thick HS Boron plate, 800 rpm, first plunge depth of 2.5 mm with a weld time of 8 s and plunge rate of 0.313 mm/s, second plunge depth of 0.2 mm with a weld time of 2 s and plunge rate of 0.1 mm/s, load ratio of 0.1 at 10 Hz	[73]
□ AA6xxx-Steel Lap Shear	1.3 mm thick 6xxx plate, 0.8 mm thick steel plate, 1500 rpm, 0.85 plunge depth, 5 s dwell time	[72]
⊞ AA6xxx-Steel Cross Tension	1.3 mm thick 6xxx plate, 0.8 mm thick steel plate, 1500 rpm, 0.85 plunge depth, 5 s dwell time	[72]
o AA5754-AA7075 Lap Shear	2.0 mm thick 5754 plate, 1.6 mm thick 7075 plate	[67]
⊕ AA5754-AA7075 Cross Tension	2.0 mm thick 5754 plate, 1.6 mm thick 7075 plate	[67]
△ AA7075-AA5754 Cross Tension	1.6 mm thick 7075 plate, 2.0 mm thick 5754 plate	[67]
◇ AA7075-AA5754 Lap Shear	1.6 mm thick 7075 plate, 2.0 mm thick 5754 plate	[67]
★ DP 590-DP 590 Lap Shear	1.0 mm thick plates, 2400 rpm, 1.6 mm plunge depth, 1 s dwell time, 3.8 mm/s plunge speed	[74]
⊕ DP 590-IF 270 Lap Shear	1.0 mm thick plates, 2400 rpm, 1.6 mm plunge depth, 1 s dwell time, 3.8 mm/s plunge speed	[74]

(Continued)

Table A.30 (Continued)		
Legend	Description	Ref.
★ IF 270-DP 590 Lap Shear	1.0 mm thick plates, 2400 rpm, 1.6 mm plunge depth, 1 s dwell time, 3.8 mm/s plunge speed	[74]
☆ IF 270-IF 270 Lap Shear	1.0 mm thick plates, 2400 rpm, 1.6 mm plunge depth, 1 s dwell time, 3.8 mm/s plunge speed	[74]
+ AA6061-AA6061 Conventional Pin Tool Lap Shear	2 mm thick plates, 3000 rpm, 1 mm plunge depth, 10 mm/min plunge speed, $R = 0.1$ at 10 Hz	[68]
✳ AA6061-AA6061 No Pin Tool Lap Shear	2 mm thick plates, 3000 rpm, 1 mm plunge depth, 10 mm/min plunge speed, $R = 0.1$ at 10 Hz	[68]
◗ AA6061-SPCC Steel, No Pin Tool Lap Shear	2 mm thick plates, 3000 rpm, 1 mm plunge depth, 10 mm/min plunge speed, $R = 0.1$ at 10 Hz	[68]

A.4.3 FSSW Steels

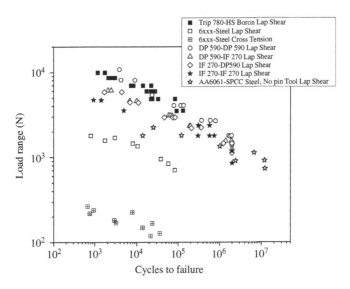

Figure A.31 Fatigue data for FSSW of steels.

Table A.31 Description of Data in Fig. A.31

Legend	Description	Ref.
■ Trip 780-HS Boron Lap Shear	1.5 mm thick Trip 780 plate, 1.4 mm thick HS Boron plate, 800 rpm, first plunge depth of 2.5 mm with a weld time of 8 s and plunge rate of 0.313 mm/s, second plunge depth of 0.2 mm with a weld time of 2 s and plunge rate of 0.1 mm/s, load ratio of 0.1 at 10 Hz	[73]
□ AA6xxx-Steel Lap Shear	1.3 mm thick 6xxx plate, 0.8 mm thick steel plate, 1500 rpm, 0.85 plunge depth, 5 s dwell time	[72]
⊞ AA6xxx-Steel Cross Tension	1.3 mm thick 6xxx plate, 0.8 mm thick steel plate, 1500 rpm, 0.85 plunge depth, 5 s dwell time	[72]
o DP 590-DP 590 Lap Shear	1.0 mm thick plates, 2400 rpm, 1.6 mm plunge depth, 1 s dwell time, 3.8 mm/s plunge speed	[74]
△ DP 590-IF 270 Lap Shear	1.0 mm thick plates, 2400 rpm, 1.6 mm plunge depth, 1 s dwell time, 3.8 mm/s plunge speed	[74]
◇ IF 270-DP 590 Lap Shear	1.0 mm thick plates, 2400 rpm, 1.6 mm plunge depth, 1 s dwell time, 3.8 mm/s plunge speed	[74]
★ IF 270-IF 270 Lap Shear	1.0 mm thick plates, 2400 rpm, 1.6 mm plunge depth, 1 s dwell time, 3.8 mm/s plunge speed	[74]
☆ AA6061-SPCC Steel, No Pin Tool Lap Shear	2 mm thick plates, 3000 rpm, 4 s dwell time, 1 mm plunge depth, 10 mm/min plunge speed, $R = 0.1$ at 10 Hz	[68]

A.4.4 FSSW Magnesium Alloys

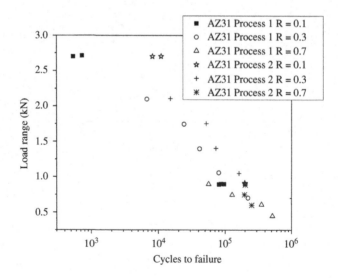

Figure A.32 Fatigue data for FSSW of magnesium alloys.

Table A.32 Description of Data in Fig. A.32

Data Set	Description	Ref.
■ AZ31 Process 1 $R = 0.1$	2 mm thick sheet, 1000 rpm, 20 mm/min, 0.5 mm shoulder depth, 2.5 s dwell time, tested at $R = 0.1$	[75]
○ AZ31 Process 1 $R = 0.3$	2 mm thick sheet, 1000 rpm, 20 mm/min, 0.5 mm shoulder depth, 2.5 s dwell time, tested at $R = 0.3$	[75]
△ AZ31 Process 1 $R = 0.7$	2 mm thick sheet, 1000 rpm, 20 mm/min, 0.5 mm shoulder depth, 2.5 s dwell time, tested at $R = 0.7$	[75]
☆ AZ31 Process 2 $R = 0.1$	2 mm thick sheet, 750 rpm, 20 mm/min, 0.1 mm shoulder depth, 2.5 s dwell time, tested at $R = 0.1$	[75]
+ AZ31 Process 2 $R = 0.3$	2 mm thick sheet, 750 rpm, 20 mm/min, 0.1 mm shoulder depth, 2.5 s dwell time, tested at $R = 0.3$	[75]
✳ AZ31 Process 2 $R = 0.7$	2 mm thick sheet, 750 rpm, 20 mm/min, 0.1 mm shoulder depth, 2.5 s dwell time, tested at $R = 0.7$	[75]

A.4.5 FSLW Magnesium Alloys

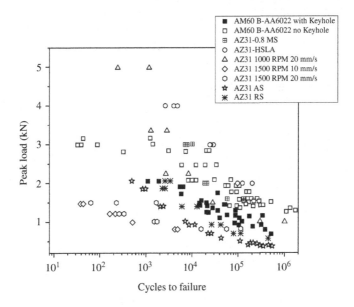

Figure A.33 Fatigue data for FSLW of magnesium.

Table A.33 Description of Data in Fig. A.33

Data Set	Description	Ref.
■ AM60 B-AA6022 with Keyhole	3.1 mm thick super vacuum die-cast AM60 B to 1.5 mm thick rolled AA6022-T4 friction stir linear welded at 1500 rpm and 75 mm/min with keyhole feature left unfilled	[76]
□ AM60 B-AA6022 no Keyhole	3.1 mm thick super vacuum die-cast AM60 B to 1.5 mm thick rolled AA6022-T4 friction stir linear welded at 1500 rpm and 75 mm/min with keyhole feature filled in	[76]
⊞ AZ31-0.8 MS	2.33 mm thick AZ31-O sheet friction stir linear welded to 0.8 mm thick MS sheet at 500 rpm and 1.67 mm/s	[77]
○ AZ31-HSLA	2.33 mm thick AZ31-O sheet friction stir linear welded to 1.5 mm thick HSLA sheet at 500 rpm and 1.67 mm/s	[77]
△ AZ31 1000 rpm 20 mm/s	2 mm thick AZ31B-H24 friction stir linear welded at 1000 rpm and 20 mm/s	[78]
◇ AZ31 1500 rpm 10 mm/s	2 mm thick AZ31B-H24 friction stir linear welded at 1500 rpm and 10 mm/s	[78]
○ AZ31 1500 rpm 20 mm/s	2 mm thick AZ31B-H24 friction stir linear welded at 1500 rpm and 20 mm/s	[78]
☆ AZ31 As	2 mm thick AZ31 sheet friction stir linear welded at 2000 rpm and 4.6 mm/s loaded in an advancing side configuration	[79]
✳ AZ31 RS	2 mm thick AZ31 sheet friction stir linear welded at 2000 rpm and 4.6 mm/s loaded in a retreating side configuration	[79]

A.4.6 FSLW Dissimilar Metals

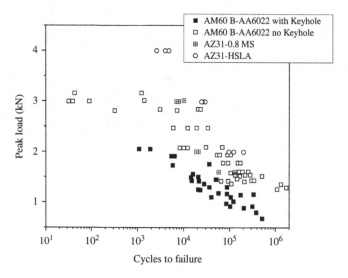

Figure A.34 Fatigue data for overlapped welds of dissimilar metals.

Table A.34 Description of Data in Fig. A.34		
Data Set	Description	Ref.
■ AM60 B-AA6022 with Keyhole	3.1 mm thick super vacuum die-cast AM60 B to 1.5 mm thick rolled AA6022-T4 friction stir linear welded at 1500 rpm and 75 mm/min with keyhole feature left unfilled	[76]
□ AM60 B-AA6022 no Keyhole	3.1 mm thick super vacuum die-cast AM60 B to 1.5 mm thick rolled AA6022-T4 friction stir linear welded at 1500 rpm and 75 mm/min with keyhole feature filled in	[76]
⊞ AZ31-0.8 MS	2.33 mm thick AZ31-O sheet friction stir linear welded to 0.8 mm thick MS sheet at 500 rpm and 1.67 mm/s	[77]
o AZ31-HSLA	2.33 mm thick AZ31-O sheet friction stir linear welded to 1.5 mm thick HSLA sheet at 500 rpm and 1.67 mm/s	[77]

REFERENCES

[1] Sun G, Niu J, Wang D, Chen S. Fatigue experimental analysis and numerical simulation of FSW joints for 2219 Al-Cu alloy. Fatigue Fract Eng Mater Struct 2015;38:445–55. Available from: https://doi.org/10.1111/ffe.12245.

[2] Ni DR, Chen DL, Xiao BL, Wang D, Ma ZY. Residual stresses and high cycle fatigue properties of friction stir welded SiCp/AA2009 composites. Int J Fatigue 2013;55:64–73. Available from: https://doi.org/10.1016/j.ijfatigue.2013.05.010.

[3] Cavaliere P, Cabibbo M, Panella F, Squillace A. 2198 Al-Li plates joined by friction stir welding: mechanical and microstructural behavior. Mater Des 2009;30:3622–31. Available from: https://doi.org/10.1016/j.matdes.2009.02.021.

[4] Boni L, Lanciotti A, Polese C. "Size effect" in the fatigue behavior of friction stir welded plates. Int J Fatigue 2015;80:238–45. Available from: https://doi.org/10.1016/j.ijfatigue.2015.06.010.

[5] Malarvizhi S, Balasubramanian V. Effects of welding processes and post-weld aging treatment on fatigue behavior of AA2219 aluminium alloy joints. J Mater Eng Perform 2011;20:359–67. Available from: https://doi.org/10.1007/s11665-010-9682-5.

[6] Cai B, Zheng ZQ, He DQ, Li SC, Li HP. Friction stir weld of 2060 Al-Cu-Li alloy: microstructure and mechanical properties. J Alloys Compd 2015;649:19–27. Available from: https://doi.org/10.1016/j.jallcom.2015.02.124.

[7] Tavares SMO, dos Santos JF, de Castro PMST. Friction stir welded joints of Al-Li Alloys for aeronautical applications: butt-joints and tailor welded blanks. Theor Appl Fract Mech 2013;65:8–13. Available from: https://doi.org/10.1016/j.tafmec.2013.05.002.

[8] Wu M, Wu CS, Gao S. Effect of ultrasonic vibration on fatigue performance of AA 2024-T3 friction stir weld joints. J Manuf Process 2017;29:85–95. Available from: https://doi.org/10.1016/j.jmapro.2017.07.023.

[9] Cavaliere P, De Santis A, Panella F, Squillace A. Effect of welding parameters on mechanical and microstructural properties of dissimilar AA6082-AA2024 joints produced by friction stir welding. Mater Des 2009;30:609–16. Available from: https://doi.org/10.1016/j.matdes.2008.05.044.

[10] Ghidini T, Dalle Donne C. Fatigue life predictions using fracture mechanics methods. Eng Fract Mech 2009;76:134–48. Available from: https://doi.org/10.1016/j.engfracmech.2008.07.008.

[11] Franchim AS, Fernandez FF, Travessa DN. Microstructural aspects and mechanical properties of friction stir welded AA2024-T3 aluminium alloy sheet. Mater Des 2011;32:4684−8. Available from: https://doi.org/10.1016/j.matdes.2011.06.055.

[12] Uematsu Y, Tokaji K, Shibata H, Tozaki Y, Ohmune T. Fatigue behaviour of friction stir welds without neither welding flash nor flaw in several aluminium alloys. Int J Fatigue 2009;31:1443−53. Available from: https://doi.org/10.1016/j.ijfatigue.2009.06.015.

[13] Jesus JS, Costa JM, Loureiro A, Ferreira JM. Fatigue strength improvement of GMAW T-welds in AA 5083 by friction-stir processing. Int J Fatigue 2017;97:124−34. Available from: https://doi.org/10.1016/j.ijfatigue.2016.12.034.

[14] Grujicic M, Arakere G, Pandurangan B, Hariharan A, Yen C-F, Cheeseman BA, et al. Statistical analysis of high-cycle fatigue behavior of friction stir welded AA5083-H321. J Mater Eng Perform 2011;20:855−64. Available from: https://doi.org/10.1007/s11665-010-9725-y.

[15] Feistauer EE, Bergmann LA, Barreto LS, dos Santos JF. Mechanical behaviour of dissimilar friction stir welded tailor welded blanks in Al-Mg alloys for marine applications. Mater Des 2014;59:323−32. Available from: https://doi.org/10.1016/j.matdes.2014.02.042.

[16] Sano Y, Masaki K, Gushi T, Sano T. Improvement in fatigue performance of friction stir welded A6061-T6 aluminum alloy by laser peening without coating. Mater Des 2012;36:809−14. Available from: https://doi.org/10.1016/j.matdes.2011.10.053.

[17] Moreira PMGP, de Oliveira FMF, de Castro PMST. Fatigue behaviour of notched specimens of friction stir welded aluminium alloy 6063-T6. J Mater Process Technol 2008;207:283−92. Available from: https://doi.org/10.1016/j.jmatprotec.2007.12.113.

[18] Minak G, Ceschini L, Boromei I, Ponte M. Fatigue properties of friction stir welded particulate reinforced aluminium matrix composites. Int J Fatigue 2010;32:218−26. Available from: https://doi.org/10.1016/j.ijfatigue.2009.02.018.

[19] Abdulstaar MA, Al-Fadhalah KJ, Wagner L. Microstructural variation through weld thickness and mechanical properties of peened friction stir welded 6061 aluminum alloy joints. Mater Charact 2017;126:64−73. Available from: https://doi.org/10.1016/j.matchar.2017.02.011.

[20] da Silva J, Costa JM, Loureiro A, Ferreira JM. Fatigue behaviour of AA6082-T6 MIG welded butt joints improved by friction stir processing. Mater Des 2013;51:315−22. Available from: https://doi.org/10.1016/j.matdes.2013.04.026.

[21] Cavaliere P, Santis AD, Panella F, Squillace A. Thermoelasticity and CCD analysis of crack propagation in AA6082 friction stir welded joints. Int J Fatigue 2009;31:385−92. Available from: https://doi.org/10.1016/j.ijfatigue.2008.07.016.

[22] Krasnowski K, Dymek S. A comparative analysis of the impact of tool design to fatigue behavior of single-sided and double-sided welded butt joints of EN AW 6082-T6 alloy. J Mater Eng Perform 2013;22:3818−24. Available from: https://doi.org/10.1007/s11665-013-0711-z.

[23] He C, Kitamura K, Yang K, Liu YJ, Wang QY, Chen Q. Very high cycle fatigue crack initiation mechanism in nugget zone of AA 7075 friction stir welded joint. Adv Mater Sci Eng 2017;2017:7189369. Available from: https://doi.org/10.1155/2017/7189369.

[24] Deng C, Wang H, Gong B, Li X, Lei Z. Effects of microstructural heterogeneity on very high cycle fatigue properties of 7050-T7451 aluminum alloy friction stir butt welds. Int J Fatigue 2016;83:100−8. Available from: https://doi.org/10.1016/j.ijfatigue.2015.10.001.

[25] Deng C, Gao R, Gong B, Yin T, Liu Y. Correlation between micro-mechanical property and very high cycle fatigue (VHCF) crack initiation in friction stir welds of 7050 aluminum alloy. Int J Fatigue 2017;104:283−92. Available from: https://doi.org/10.1016/j.ijfatigue.2017.07.028.

[26] Gemme F, Verreman Y, Dubourg L, Wanjara P. Effect of welding parameters on microstructure and mechanical properties of AA7075-T6 friction stir welded joints. Fatigue Fract Eng Mater Struct 2011;34:877−86. Available from: https://doi.org/10.1111/j.1460-2695.2011.01580.x.

[27] Nelaturu P, Jana S, Mishra RS, Grant G, Carlson BE. Influence of friction stir processing on the room temperature fatigue cracking mechanisms of A356 aluminum alloy. Mater Sci Eng A 2018;716:165−78. Available from: https://doi.org/10.1016/j.msea.2018.01.044.

[28] Tajiri A, Uematsu Y, Kakiuchi T, Tozaki Y, Suzuki Y, Afrinaldi A. Effect of friction stir processing conditions on fatigue behavior and texture development in A356-T6 cast aluminum alloy. Int J Fatigue 2015;80:192−202. Available from: https://doi.org/10.1016/j.ijfatigue.2015.06.001.

[29] Sahandi Zangabad P, Khodabakhshi F, Simchi A, Kokabi AH. Fatigue fracture of friction-stir processed Al-Al3Ti-MgO hybrid nanocomposites. Int J Fatigue 2016;87:266−78. Available from: https://doi.org/10.1016/j.ijfatigue.2016.02.007.

[30] Jana S, Mishra RS, Baumann JB, Grant G. Effect of friction stir processing on fatigue behavior of an investment cast Al-7Si-0.6 Mg alloy. Acta Mater 2010;58:989−1003. Available from: https://doi.org/10.1016/j.actamat.2009.10.015.

[31] Sahu PK, Pal S. Mechanical properties of dissimilar thickness aluminium alloy weld by single/double pass FSW. J Mater Process Technol 2017;243:442−55. Available from: https://doi.org/10.1016/j.jmatprotec.2017.01.009.

[32] Lertora E, Gambaro C. AA8090 Al-Li Alloy FSW parameters to minimize defects and increase fatigue life. Int J Mater Form 2010;3:1003−6. Available from: https://doi.org/10.1007/s12289-010-0939-1.

[33] Jana S, Mishra RS, Baumann JB, Grant G. Effect of stress ratio on the fatigue behavior of a friction stir processed cast Al-Si-Mg alloy. Scr Mater 2009;61:992−5. Available from: https://doi.org/10.1016/j.scriptamat.2009.08.011.

[34] Zhou L, Li ZY, Nakata K, Feng JC, Huang YX, Liao JS. Microstructure and fatigue behavior of friction stir-welded noncombustive Mg-9Al-Zn-Ca magnesium alloy. J Mater Eng Perform 2016;25:2403−11. Available from: https://doi.org/10.1007/s11665-016-2061-0.

[35] Padmanaban G, Balasubramanian V. Fatigue performance of pulsed current gas tungsten arc, friction stir and laser beam welded AZ31B magnesium alloy joints. Mater Des 2010;31:3724−32. Available from: https://doi.org/10.1016/j.matdes.2010.03.013.

[36] Ni DR, Wang D, Feng AH, Yao G, Ma ZY. Enhancing the high-cycle fatigue strength of Mg-9Al-1Zn casting by friction stir processing. Scr Mater 2009;61:568−71. Available from: https://doi.org/10.1016/j.scriptamat.2009.05.023.

[37] Yang J, Ni DR, Xiao BL, Ma ZY. Non-uniform deformation in a friction stir welded Mg-Al-Zn joint during stress fatigue. Int J Fatigue 2014;59:9−13. Available from: https://doi.org/10.1016/j.ijfatigue.2013.10.004.

[38] Ito K, Okuda T, Ueji R, Fujii H, Shiga C. Increase of bending fatigue resistance for tungsten inert gas welded SS400 steel plates using friction stir processing. Mater Des 2014;61:275−80. Available from: https://doi.org/10.1016/j.matdes.2014.04.076.

[39] Lakshminarayanan AK, Balasubramanian V. Assessment of fatigue life and crack growth resistance of friction stir welded AISI 409M ferritic stainless steel joints. Mater Sci Eng A 2012;539:143−53. Available from: https://doi.org/10.1016/j.msea.2012.01.071.

[40] Edwards P, Ramulu M. Identification of process parameters for friction stir welding Ti-6Al-4V. J Eng Mater Technol 2010;132:031006. Available from: https://doi.org/10.1115/1.4001302.

[41] Xu WF, Liu JH, Chen DL, Luan GH. Low-cycle fatigue of a friction stir welded 2219-T62 aluminum alloy at different welding parameters and cooling conditions. Int J Adv Manuf Technol 2014;74:209−18. Available from: https://doi.org/10.1007/s00170-014-5988-z.

[42] Xu WF, Liu JH, Chen DL, Luan GH, Yao JS. Change of microstructure and cyclic deformation behavior along the thickness in a friction-stir-welded aluminum alloy. Scr Mater 2012;66:5–8. Available from: https://doi.org/10.1016/j.scriptamat.2011.08.021.

[43] Rodriguez RI, Jordon JB, Allison PG, Rushing T, Garcia L. Low-cycle fatigue of dissimilar friction stir welded aluminum alloys. Mater Sci Eng A 2016;654:236–48. Available from: https://doi.org/10.1016/j.msea.2015.11.075.

[44] Feng AH, Chen DL, Ma ZY. Microstructure and low-cycle fatigue of a friction-stir-welded 6061 aluminum alloy. Metall Mater Trans A Phys Metall Mater Sci 2010;41:2626–41. Available from: https://doi.org/10.1007/s11661-010-0279-2.

[45] Cisko AR, Jordon B, Rodriguez R, Rao H, Allison PG. Microstructure-sensitive fatigue modeling of friction stir welded aluminum alloy 6061. ASME 2015 International Mechanical Engineering Congress and Exposition, Volume 14: Emerging Technologies; Safety Engineering and Risk Analysis; Materials: Genetics to Structures. Houston, TX: ASME; 2015, p. V014T11A003. https://doi.org/10.1115/IMECE2015-52307.

[46] Rodriguez RI, Jordon JB, Allison PG, Rushing T, Garcia L. Corrosion effects on the fatigue behavior of dissimilar friction stir welding of high-strength aluminum. Mater Sci Eng A 2018;742:255–68.

[47] Ni DR, Chen DL, Yang J, Ma ZY. Low cycle fatigue properties of friction stir welded joints of a semi-solid processed AZ91D magnesium alloy. Mater Des 2014;56:1–8. Available from: https://doi.org/10.1016/j.matdes.2013.10.081.

[48] Yang J, Ni DR, Wang D, Xiao BL, Ma ZY. Strain-controlled low-cycle fatigue behavior of friction stir-welded AZ31 magnesium alloy. Metall Mater Trans A 2014;45:2101–15. Available from: https://doi.org/10.1007/s11661-013-2129-5.

[49] Sun GQ, Niu JP, Chen YJ, Sun FY, Shang DG, Chen SJ. Experimental research on fatigue failure for 2219-T6 aluminum alloy friction stir-welded joints. J Mater Eng Perform 2017;26:3767–74. Available from: https://doi.org/10.1007/s11665-017-2836-y.

[50] Malarvizhi S, Balasubramanian V. Fatigue crack growth resistance of gas tungsten arc, electron beam and friction stir welded joints of AA2219 aluminium alloy. Mater Des 2011;32:1205–14. Available from: https://doi.org/10.1016/j.matdes.2010.10.019.

[51] Wang W, Qiao K, Wu JL, Li TQ, Cai J, Wang KS. Fatigue properties of friction stir welded joint of ultrafine-grained 2024 aluminium alloy. Sci Technol Weld Join 2017;22:110–19. Available from: https://doi.org/10.1080/13621718.2016.1203177.

[52] Milan MT, Bose WW, Tarpani JR. Fatigue crack growth behavior of friction stir welded 2024-T3 aluminum alloy tested under accelerated salt fog exposure. Mater Perform Charact 2014;3. Available from: https://doi.org/10.1520/MPC20130036 20130036.

[53] Moreira PMGP, de Jesus AMP, de Figueiredo MAV, Windisch M, Sinnema G, de Castro PMST. Fatigue and fracture behaviour of friction stir welded aluminium-lithium 2195. Theor Appl Fract Mech 2012;60:1–9. Available from: https://doi.org/10.1016/j.tafmec.2012.06.001.

[54] Ma YE, Staron P, Fischer T, Irving PE. Size effects on residual stress and fatigue crack growth in friction stir welded 2195-T8 aluminium—part I: experiments. Int J Fatigue 2011;33:1417–25. Available from: https://doi.org/10.1016/j.ijfatigue.2011.05.006.

[55] Hatamleh O, Hill M, Forth S, Garcia D. Fatigue crack growth performance of peened friction stir welded 2195 aluminum alloy joints at elevated and cryogenic temperatures. Mater Sci Eng A 2009;519:61–9. Available from: https://doi.org/10.1016/j.msea.2009.04.049.

[56] Cavaliere P, De Santis A, Panella F, Squillace A. Effect of anisotropy on fatigue properties of 2198 Al-Li plates joined by friction stir welding. Eng Fail Anal 2009;16:1856–65. Available from: https://doi.org/10.1016/j.engfailanal.2008.09.024.

[57] Ma YE, Irving P. Residual stress effects and fatigue behavior of friction-stir-welded 2198-T8 Al-Li alloy joints. J Aircr 2011;48:1238–44. Available from: https://doi.org/10.2514/1. c031242.

[58] Fonda RW, Pao PS, Jones HN, Feng CR, Connolly BJ, Davenport AJ. Microstructure, mechanical properties, and corrosion of friction stir welded Al 5456. Mater Sci Eng A 2009;519:1–8. Available from: https://doi.org/10.1016/j.msea.2009.04.034.

[59] Tra TH, Okazaki M, Suzuki K. Fatigue crack propagation behavior in friction stir welding of AA6063-T5: roles of residual stress and microstructure. Int J Fatigue 2012;43:23–9. Available from: https://doi.org/10.1016/j.ijfatigue.2012.02.003.

[60] Pirondi A, Collini L. Analysis of crack propagation resistance of Al-Al2O3 particulate-reinforced composite friction stir welded butt joints. Int J Fatigue 2009;31:111–21. Available from: https://doi.org/10.1016/j.ijfatigue.2008.05.003.

[61] D'Urso G, Giardini C, Lorenzi S, Pastore T. Fatigue crack growth in the welding nugget of FSW joints of a 6060 aluminum alloy. J Mater Process Technol 2014;214:2075–84. Available from: https://doi.org/10.1016/j.jmatprotec.2014.01.013.

[62] Pirondi A, Collini L, Fersini D. Fracture and fatigue crack growth behaviour of PMMC friction stir welded butt joints. Eng Fract Mech 2008;75:4333–42. Available from: https:// doi.org/10.1016/j.engfracmech.2008.05.001.

[63] Hatamleh O, Forth S, Reynolds AP. Fatigue crack growth of peened friction stir-welded 7075 aluminum alloy under different load ratios. J Mater Eng Perform 2010;19:99–106. Available from: https://doi.org/10.1007/s11665-009-9439-1.

[64] Edwards PD, Ramulu M. Comparative study of fatigue and fracture in friction stir and electron beam welds of 24 mm thick titanium alloy Ti-6Al-4 V. Fatigue Fract Eng Mater Struct 2016;39:1226–40. Available from: https://doi.org/10.1111/ffe.12434.

[65] Muzvidziwa M, Okazaki M, Suzuki K, Hirano S. Role of microstructure on the fatigue crack propagation behavior of a friction stir welded Ti-6Al-4V. Mater Sci Eng A 2016;652:59–68. Available from: https://doi.org/10.1016/j.msea.2015.11.065.

[66] Tran VX, Pan J, Pan T. Fatigue behavior of aluminum 5754-O and 6111-T4 spot friction welds in lap-shear specimens. Int J Fatigue 2008;30:2175–90. Available from: https://doi. org/10.1016/j.ijfatigue.2008.05.025.

[67] Tran VX, Pan J, Pan T. Fatigue behavior of spot friction welds in lap-shear and cross-tension specimens of dissimilar aluminum sheets. Int J Fatigue 2010;32:1022–41. Available from: https://doi.org/10.1016/j.ijfatigue.2009.11.009.

[68] Uematsu Y, Tokaji K, Tozaki Y, Nakashima Y, Shimizu T. Fatigue behaviour of dissimilar friction stir spot welds between A6061-T6 and low carbon steel sheets welded by a scroll grooved tool without probe. Fatigue Fract Eng Mater Struct 2011;34:581–91. Available from: https://doi.org/10.1111/j.1460-2695.2010.01549.x.

[69] Uematsu Y, Tokaji K. Comparison of fatigue behaviour between resistance spot and friction stir spot welded aluminium alloy sheets. Sci Technol Weld Join 2009;14:62–71. Available from: https://doi.org/10.1179/136217108X338908.

[70] Su ZM, He RY, Lin PC, Dong K. Fatigue of alclad AA2024-T3 swept friction stir spot welds in cross-tension specimens. J Mater Process Technol 2016;236:162–75. Available from: https://doi.org/10.1016/j.jmatprotec.2016.05.014.

[71] Lin PC, Lo SM, Wu SP. Fatigue life estimations of alclad AA2024-T3 friction stir clinch joints. Int J Fatigue 2018;107:13–26. Available from: https://doi.org/10.1016/j. ijfatigue.2017.10.011.

[72] Tran V-X, Pan J. Fatigue behavior of dissimilar spot friction welds in lap-shear and cross-tension specimens of aluminum and steel sheets. Int J Fatigue 2010;32:1167–79. Available from: https://doi.org/10.1016/J.IJFATIGUE.2009.12.011.

[73] Hong SH, Sung S-J, Pan J. Failure mode and fatigue behavior of dissimilar friction stir spot welds in lap-shear specimens of transformation-induced plasticity steel and hot-stamped boron steel sheets. J Manuf Sci Eng 2015;137:051023. Available from: https://doi.org/10.1115/1.4031235.

[74] Sarkar R, Sengupta S, Pal TK, Shome M. Microstructure and mechanical properties of friction stir spot-welded if/dp dissimilar steel joints. Metall Mater Trans A Phys Metall Mater Sci 2015;46:5182–200. Available from: https://doi.org/10.1007/s11661-015-3116-9.

[75] Rao HM, Jordon JB, Barkey ME, Guo YB, Su X, Badarinarayan H. Influence of structural integrity on fatigue behavior of friction stir spot welded AZ31 Mg alloy. Mater Sci Eng A 2013;564:369–80. Available from: https://doi.org/10.1016/J.MSEA.2012.11.076.

[76] Rao HM, Jordon JB, Boorgu SK, Kang H, Yuan W, Su X. Influence of the key-hole on fatigue life in friction stir linear welded aluminum to magnesium. Int J Fatigue 2017;105:16–26. Available from: https://doi.org/10.1016/j.ijfatigue.2017.08.012.

[77] Jana S, Hovanski Y. Fatigue behaviour of magnesium to steel dissimilar friction stir lap joints. Sci Technol Weld Join 2012;17:141–5. Available from: https://doi.org/10.1179/1362171811Y.0000000083.

[78] Naik BS, Chen DL, Cao X, Wanjara P. Microstructure and fatigue properties of a friction stir lap welded magnesium alloy. Metall Mater Trans A Phys Metall Mater Sci 2013;44:3732–46. Available from: https://doi.org/10.1007/s11661-013-1728-5.

[79] Moraes JFC, Rodriguez RI, Jordon JB, Su X. Effect of overlap orientation on fatigue behavior in friction stir linear welds of magnesium alloy sheets. Int J Fatigue 2017;100:1–11. Available from: https://doi.org/10.1016/j.ijfatigue.2017.02.018.

Printed in the United States
By Bookmasters